Plant Engineering magazine's

FLUID POWER

Handbook

Volume 1: System Design, Maintenance, and Troubleshooting

Plant Engineering magazine's

FLUID POWER

Handbook

Volume 1: System Design, Maintenance, and Troubleshooting

Anton H. Hehn

Gulf Publishing Company
Houston, London, Paris, Zurich, Tokyo

Plant
Engineering magazine's
FLUID POWER
Handbook

Volume 1: System Design, Maintenance, and Troubleshooting

Gulf Publishing Company
Book Division
P.O. Box 2608, Houston, Texas 77252-2608

10   9   8   7   6   5   4   3   2   1

Library of Congress Cataloging-in-Publication Data

Hehn, Anton H., 1937–
    Plant engineering's fluid power handbook / Anton H. Hehn.
        p.   cm.
    Includes index.
    Contents: v. 1. System design, maintenance, and troubleshooting.
    ISBN 0-88415-072-0
    1. Fluid power technology—Handbooks, manuals, etc.   2. Plant engineering—Handbooks, manuals, etc.   I. Title.
    TJ840.H44   1993
    620.1'06—dc20   92-46504
                                                          CIP

Printed in the United States of America.

Printed on Acid-Free (∞) paper.

Contents

Preface

Fluid power is a relatively young field of energy transmission and control. The manufacture of hydraulic and pneumatic equipment for typical industrial applications has grown very rapidly since the mid-1940s. Present-day hydraulic and pneumatic systems can economically convert mechanical energy into fluid energy, and this energy can be regulated to achieve direction, speed, and force control. There is no other type of power transmission method that provides the range of control for force, speed, and direction as is possible with fluid power. Today, fluid power equipment is used in nearly every branch of industry, including plastics processing, packaging, pharmaceuticals, petrochemicals, materials handling, construction, machine tools, and others. This ever-increasing use of fluid power has resulted in an expanded need for engineers, managers, technicians, and mechanics who are familiar with the operation, use, and care of this equipment.

This two-volume set of handbooks has been prepared to provide plant engineers, designers, and mechanics with a reference source in the area of fluid power. Volume 1 describes the basic principles of hydraulic and pneumatic systems; how the components are used and how they function; and how to maintain and troubleshoot fluid power systems. Volume 2 deals with the design and application aspects of hydraulic and pneumatic systems. Included here are component sizing and selection procedures, hydraulic circuit design, and applications of hydraulic and pneumatic systems. Also, electrohydraulic proportional valves, servovalves, and servosystems are included here. The material is practical in character; theory is kept to a minimum. These books provide the reader with answers to many of the most common questions pertaining to the operation, application, maintenance, and troubleshooting of fluid power systems.

The information presented is based on many years of designing, maintaining, and troubleshooting equipment powered and controlled by fluid power. Much of the material stems from the *Plant Engineering* magazine library of articles dealing with this subject.

Great thanks are due to the editors of *Plant Engineering* for their assistance with this project. In particular, wholehearted thanks to

Richard L. Dunn and Joseph L. Foszcz for their help and guidance. Many thanks also to the fluid power equipment manufacturers for their generous contributions of technical information and illustrations.

Anton H. Hehn
Michigan City, Indiana

Plant Engineering magazine's

FLUID POWER

Handbook

Volume 1: System Design, Maintenance, and Troubleshooting

Introduction to Fluid Power

Why Fluid Power?

In the 300 years since the experiments of Mariotte, Torricelli, and Bernoulli laid the foundation for the development of the science of fluid power, it has become a significant factor in plants ranging from huge automobile assembly facilities to repair shops. Fluid power encompasses the utilization of liquids (oil, water, and synthetics) and gases (normally compressed air) as media for the transmission of energy. In this service, considerable adaptability is inherent in fluids, permitting simple and efficient conversion from one form of energy to another, from one type of motion to another, and from a given force to an increased or reduced force, with complete and very accurate control.

- Fluid motors can be stalled without damage. Many can be suddenly reversed by valving pressure from the inlet port to the outlet port.
- The size-to-horsepower (hp) and weight-to-horsepower ratios of fluid motors are very favorable. For instance, a 50-hp hydraulic motor weighs approximately 78 lb, whereas a 50-hp electric motor weighs approximately 600 lb.
- Cylinders can be stalled without damage. This advantage is especially important in applications in which jamming may occur.
- The piston of a hydraulic or an air-hydraulic cylinder can be stopped at intermediate points on the advancing or retracting stroke with a good degree of accuracy.
- Piston speeds can be controlled to produce a relatively fine feed or a high speed. (Hydraulic cylinders are the easier of the two to control.) Skip feeds are practical.
- The pressure to the piston can be controlled—from just enough to cause it to move to the maximum for which the cylinder is rated. The pressure applied to the rod-end side of the piston can differ from that applied to the blind end.

1

- A fast-moving piston can be brought to a stop at the end of its stroke with relatively little shock by means of long cushions or shock absorbers.
- Power supplies can be portable, with an internal combustion engine used as the prime mover. This setup eliminates the need for electric power in areas where new construction is under way.
- Directional control valves are relatively simple in design, yet are available in many configurations—especially hydraulic valves. A large number of actuators are also available.
- Functional controls provide capability for a wide variety of motions, pressures, and forces.

Yet fluid power is far from perfect. Its drawbacks rule it out for some applications and limit its usefulness in others.

- Temperature can cause problems. For example, an air hoist or an air-operated loading platform may be troublesome in cold weather because of moisture freezing in the compressed air lines or the cylinder. Hydraulic systems in extremely cold locations often require a great deal of heat input to the controls and the pumping system.
- External leakage is common and undesirable. Hydraulic leaks are messy and create fire hazzards and dangerous conditions such as slippery floors. Compressed air leaks often go unnoticed for long periods. If the machinery were stopped in most large plants using air-operated equipment, the noise from air leaks would be incredible.
- Contamination, whether from external or internal sources, has a bad effect on the components of fluid-power systems.
- Synchronization of the movement of two or more cylinders is difficult. With air cylinders, it is almost impossible. Hydraulic cylinders can be synchronized fairly accurately with flow dividers, but one of the best methods is to use air-hydraulic cylinders and criss-cross the lines of the closed hydraulic systems — one cylinder cannot produce motion without the other cylinder providing the same motion.

Comparison of an Air System to a Hydraulic System

On many applications it will be immediately apparent which medium, air or hydraulics, is more suitable. But some jobs could be done with either medium, or with vacuum, and the following points should be considered to help make the best decision (also see Table 1.1).

Table 1.1
System Comparisons: Air—Electric—Hydraulic

Air	Electric	Hydraulic
Variable speed	Fixed speed	Variable speed
Runs cool	Runs hot	Needs cooler
Explosion proof	Explosion proofing adds to cost	Oil must be nonflammable for explosion proofing
Compact	Less compact	Most compact
Can be stalled indefinitely	Cannot be stalled	Can be stalled indefinitely
Easily reversed	Reversing limited by heat build-up	Easily reversed
Low efficiency, 20%	High efficiency, 60%	High efficiency, 50%
Low torque	High torque	High torque
Low cost	Moderate cost	High cost
Clean leaks	Shock hazard	Messy leaks

Air—Low Cost

Power Level

Branch air circuits usually operate in the ¼- to 1½-hp range, while most hydraulic systems operate from 1½-hp up. This is a general rule, as there are higher powered air systems and lower powered hydraulic systems. But a 100-hp air compressor in a large plant is usually feeding numerous branch circuits that are operating independently of one another, and a 100-hp hydraulic system is usually operating only one machine, although this machine may have several branch circuits dependent on one another.

Noise Level

If the air exhaust noise is properly muffled, an air system usually operates much more quietly than a hydraulic system of the same hp. And, in between cycles, an air system is completely silent, while in a hydraulic system the pump remains rotating in an unloaded condition. Where a hydraulic system must be used and where noise must be reduced to a minimum, the pump should be operated at the minimum rpm (revolutions per minute) that will deliver sufficient oil, and the pressure in the system should also be kept as low as possible, by using cylinders of larger bore operating at lower pressures.

Cleanliness

Normally an air system is very clean, provided the air-line oiler is not feeding an excessive amount of oil into the air, later to be blown out the control-valve exhaust. A hydraulic system that is carefully designed and constructed can be clean, but there may come a time when a cylinder-rod packing may start to leak, or a line must be disconnected to replace a component, or a filter element must be replaced. At these times there is a possibility of oil being spilled around the machine.

Speed

Lightweight mechanisms can usually be operated faster with compressed air because a large volume of air can be drawn from the storage tank, for short periods, to give a very fast cylinder speed. A hydraulic system, to match the speed of an air system, might have to have a very large pump, large valving, and large plumbing, since the power is usually generated by the pump at the rate of use, with no reserve supply. Of course, accumulators can be used in a hydraulic system, but these make a low-power system expensive.

Operating Cost

A hydraulic system usually costs considerably less to operate than a compressed-air system with the same mechanical power from the cylinder, because hydraulics is more efficient, or can be. As air is compressed, it is also heated. This heat of compression radiates from the walls of the compressor, the storage tank, and the plumbing system; or it is removed by an aftercooler. This is an escape of power from the fluid system that can be minimized by good design, but the heat losses from air compression cannot be avoided or minimized.

Initial Cost

If the cost of the compressor is considered, a compressed-air system is more costly to build than a hydraulic system. But sometimes a low-power air circuit can be added to an existing compressor that has reserve capacity, and this can be done at less cost than a hydraulic system.

rigidity - need hydraulics (intermediate stopping)

Rigidity

Some applications must have rigidity in the fluid stream, and require the use of hydraulics rather than compressed air. Examples are lifts and elevators that need to be stopped at an intermediate point in the cylinder stroke to add or remove a part of their load; the feeding of a cutting tool at a slow speed; slow movement of a machine slide or other mechanism having a large area of sliding friction; press applications where two or more cylinders are attached to an unguided or inadequately guided platen, particularly where the load is not distributed evenly over all cylinders; and any system where close control of cylinder speed or position must be maintained, or where the cylinder must be accurately stopped at a precise position in mid-stroke. These and similar applications, whether high or low power, can only be done satisfactorily with hydraulics.

Hydraulics: How It Works

Why Hydraulics Is Used

Oil hydraulics today is an essential area of knowledge for anyone who has a technical interest in industrial equipment or any mobile-type vehicles. In earth-moving equipment, self-propelled vehicles, farm tractors, and industrial machinery, demonstrate brute force with very precise control through the use of hydraulic systems. The touch of a handle lifts several yards of dirt in the bucket of a loader—trenches are dug quickly without anyone lifting a shovelful of dirt—by the use of the hidden giant hydraulics.

Hydraulics is the science dealing with work performed by liquids in motion. It is taken for granted so often that we tend to forget how important the harnessing of hydraulic energy is to our daily lives. Many knowledgeable people have said that modern machinery simply would not function without hydraulics.

The science of hydraulics is divided into two distinct categories, hydrodynamics and hydrostatics. Hydrodynamics deals with power transmitted by liquids in motion, such as water turning a turbine. Hydrostatics deals with power transmitted by confined liquids under pressure. The liquid in the hydrostatic system moves or flows, but energy is transmitted primarily because the liquid is pressurized. The term *hydraulics* as used in this book refers to the hydrostatic branch of the science.

Basic Principles of Hydraulics

The basic principles of hydraulics are few and simple:

- Liquids have no shape of their own.
- Liquids are practically incompressible.
- Liquids transmit applied pressure in all directions.
- Liquids provide great increases in work force.

Liquids have no shape of their own. They acquire the shape of any container (Figure 1.1). Because of this, oil in a hydraulic system will flow in any direction and into a passage of any size or shape.

Liquids are practically incompressible. This is shown in Figure 1.2. For safety reasons, we obviously wouldn't perform the experiment shown. However, if we were to push down on the cork of the tightly sealed jar, the liquid in the jar would not compress. The jar would shatter first. (NOTE: Liquids will compress slightly under pressure, but for our purposes they are incompressible.)

Liquids transmit applied pressure in all directions. The experiment in Figure 1.2 shattered the glass jar and also showed how liquids transmit pressure—in all directions when they are compressed. This is very important in a hydraulic system. As shown in Figure 1.3, take two cylinders of the same size (1 in.^{2}) and connect them by a tube. Fill the cylinders with oil to the level shown. Place in each cylinder a piston, which rests on the columns of oil. Now press down on one cylinder with a force of 1 lb. This pressure is created throughout the system, and an equal force of 1 lb is applied to the other piston, raising it as shown.

Liquids provide great increases in work force. Now let us take two cylinders of different sizes and connect them as shown in Figure 1.4. The first cylinder has an area of 1 in.^{2}, but the second has an area of 10 in.^{2} Again use a force of 1 lb on the piston in the smaller cylinder. Once again the pressure is created throughout the system and a pressure of 1 lb/in.^{2} is exerted on the larger cylinder. Since that cylinder has a piston area of 10 in.^{2}, the total force exerted on it is 10 lb. In other words, we have a great increase in work force.

Figure 1.1. Liquids have no shape of their own.

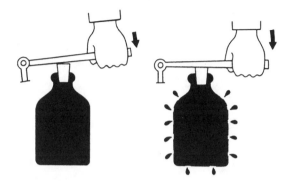

Figure 1.2. Liquids are practically incompressible.

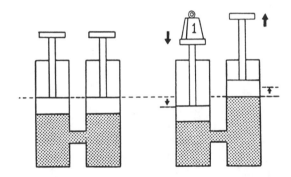

Figure 1.3. Liquids transmit applied pressure in all directions.

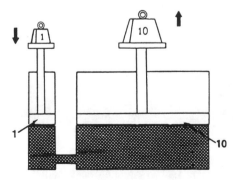

Figure 1.4. Liquids provide great increases in work force.

Pressure

In studying the basic principles of hydraulics, one must be concerned with forces, energy transfer, work, and power, and how these relate to the two fundamental conditions or phenomena encountered in a hydraulic system: pressure and flow.

Pressure and flow, of course, must be interrelated in considering work, energy, and power. On the other hand, each has its own particular job to do.

- Pressure is responsible for pushing or exerting a force or torque (Figure 1.5).
- Flow is responsible for making something move; for causing motion.

Because these jobs are often confused by the uninitiated, try to keep them distinct as we consider separately, and then together, the phenomena of pressure and flow.

What Is Pressure?

To the engineer, pressure is a term used to define how much force is exerted against a specific area. The technical definition of pressure, in fact, is force per unit area.

One example of pressure is the tendency to expand (or resistance to compression) that is present in a fluid that is being squeezed. A fluid, by definition, is any liquid or any gas (vapor).

The air that fills an automobile tire is a gas, and thus obeys the laws of fluids. When you inflate a tire, you are squeezing in more air than the tire would like to hold. The air inside the tire resists this squeezing by pushing outward on the casing of the tire. The outward push of the air is pressure.

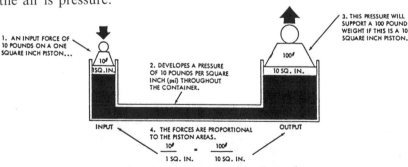

Figure 1.5.   **Pressure provides the push.**

Like all gases, air is highly compressible. That is, it can be squeezed into a smaller volume, or more air can be squeezed into the same space. As additional air is squeezed into a tire, it takes greater force, and the pressure within the tire increases.

The outward push of the air in a tire is uniform throughout. That is, all the inner surface of the tire is subject to the same amount of pressure. If it weren't, the tire would be pushed into odd shapes because of its elasticity.

Equal pressure throughout the area of confinement is a characteristic of any pressurized fluid, whether gas or liquid. The difference is that liquids are only very slightly compressible.

If you tried to force a stopper into a bottle that was completely full of water, you would experience the near-incompressibility of a liquid. Each time the stopper is pushed in, it springs back immediately when let go. And if one were so bold as to hit the stopper with a hammer, chances are the bottle would break.

When a confined liquid is pushed on, there is a pressure buildup (Figure 1.6). The pressure is still transmitted equally throughout the container. The bottle that breaks from excess pressure may break anywhere, or in several places at once.

This behavior of a fluid is what makes it possible to push it through pipes, around corners, up and down, and so on. In hydraulic systems, we use a liquid, because its near-incompressibility makes the action instantaneous, so long as the system is primed or full of liquid.

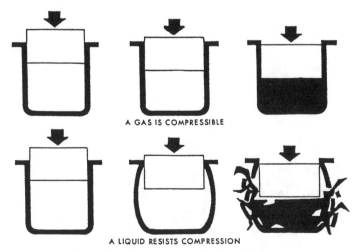

A GAS IS COMPRESSIBLE

A LIQUID RESISTS COMPRESSION

Figure 1.6. Pressure on a confined fluid.

How Pressure Is Created

It is fundamental that pressure can be created by squeezing or pushing on a confined fluid only if there is a resistance to flow. There are two ways to push on a fluid: by the action of some sort of mechanical pump, or by the force exerted by the weight of the fluid itself.

A diver cannot go to great depths in the ocean because of the tremendous pressure of the water around him. This pressure is due entirely to the force exerted by the weight of the water above the diver, and it increases in proportion to the depth. Knowing the force exerted by the weight of a cubic foot of water, we can calculate the pressure at any depth exactly.

Suppose, as shown in Figure 1.7, we isolate a column of water 1 ft² square and 10 ft high. We want to determine what the pressure is at the bottom of this column. Since the force exerted by the weight of a 1 ft³ of water is 62.4 lb and we have 10 ft³ in this example, the total force exerted by the weight of the water is 624 lb.

At the bottom, this force is distributed over 144 in.² (the equivalent of 1 ft²). Each single in.² of the bottom is subject to $^{1}/_{144}$ of the total force, or 4.33 lb. We say, then, that the pressure at this depth is 4.33 lb/in.²

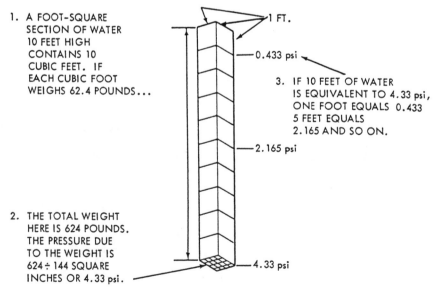

1. A FOOT-SQUARE SECTION OF WATER 10 FEET HIGH CONTAINS 10 CUBIC FEET. IF EACH CUBIC FOOT WEIGHS 62.4 POUNDS...

1 FT.

0.433 psi

3. IF 10 FEET OF WATER IS EQUIVALENT TO 4.33 psi, ONE FOOT EQUALS 0.433 5 FEET EQUALS 2.165 AND SO ON.

2.165 psi

2. THE TOTAL WEIGHT HERE IS 624 POUNDS. THE PRESSURE DUE TO THE WEIGHT IS 624 ÷ 144 SQUARE INCHES OR 4.33 psi.

4.33 psi

Figure 1.7. Pressure due to weight of fluid.

Pounds-per-square-inch is the common unit of pressure. We abbreviate it "psi." It tells us how many pounds are exerted on a unit area of 1 in.^{2}

We could just as easily create an equal pressure of 4.33 psi in a liquid as shown in Figure 1.8. If we trap the liquid under a piston that has an area of 10 in.^{2} and place a weight on the piston so that it pushes down with 43.3 lb force, a pressure of 4.33 psi will result.

Of course, it is not necessary to push downward with force exerted by the weight to create pressure in a fluid. It is only necessary to apply any kind of force. We define force as any push or pull. Weight is only one kind of force—the force of gravity on an object. We could just as easily turn the container in Figure 1.8 on its side and push on the piston with a spring or a strong right arm, or a crankshaft driven by an engine. In any case, we measure the applied force in pounds. Pressure is created in the liquid in proportion to the force.

Pascal's Law

Pascal's Law tells us this: Pressure on a confined fluid is transmitted undiminished in every direction, and acts with equal force on equal areas, and at right angles to the container walls.

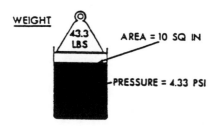

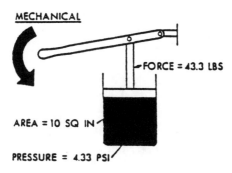

Figure 1.8. Pressure-force relationship.

We know now (1) that pressure is force per unit area, expressed as psi and, (2) that force is a push or pull, measured in pounds. In Figure 1.8, we showed what happened when we applied a force to a confined fluid through a piston. The resulting pressure in the fluid, by Pascal's Law, is equal throughout. And, every square inch of the container wall is subject to an equal force because of the pressure.

Pressure and Force Relationships

Our example of hydraulic force multiplication has given us two important relationships from Pascal's Law. We can express these relationships as equations to solve simple problems of pressure and force.

First, pressure is equal to force divided by area:

$$P = \frac{F}{A}$$

Second, the force on any area is equal to the area multiplied by the pressure on the area:

$$F = P \times A$$

When using these relationships, always use these

$$
\begin{aligned}
F \text{ (force)} &= \text{pounds} \\
P \text{ (pressure)} &= \text{psi} \\
A \text{ (area)} &= \text{square inches}
\end{aligned}
$$

If any other units are given, convert them to these units before solving the problem. For instance, convert ounces to pounds; square feet to square inches.

The diagram in Figure 1.9 makes these formulas easier to remember and simplifies problem solving. If any two values are known, the diagram shows which mathematical operation to perform to find the third by simply covering the unknown value with a finger.

For example, if force is to be determined, when the area of the actuator and pressure are known, cover the unknown "F." It can be seen that P and A must be multiplied to obtain F (Figure 1.10).

Figure 1.9.   Force, pressure, and area relationship.

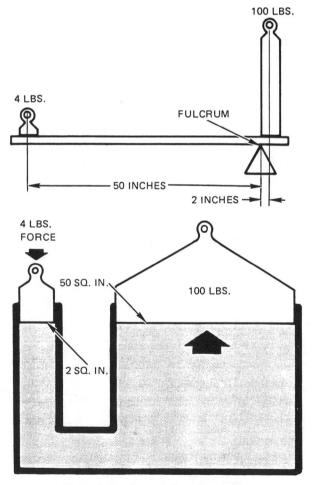

Figure 1.10.   Increase force example.

Pressure In a Parallel Connection

When several loads are connected in parallel (Figure 1.11) the oil takes the path of least resistance. Since cylinder A requires the least pressure, it will move first. Furthermore, pressure will not build up beyond the needs of A until it has reached its travel limit. Then pressure will rise just high enough to move cylinder B. Finally, when B is at its limit, pressure will rise to move cylinder C.

Flow

Flow is simply the movement of liquid caused by the difference in pressure between two points. In a hydraulic system, flow is produced by the action of a pump.

In our kitchen sink, for instance, we have atmospheric pressure. The city water works has built up a pressure or head in our pipes. When we open the tap, the pressure difference forces the water out.

Flow may be measured either by its velocity or by its flow rate. Velocity is the average speed of a liquid past a given point, measured in feet or meters per second (fps or m/sec) or feet or meters per minute (fpm or m/min). Flow rate is a measure of the volume of liquid that passes a point in a given time, usually measured in gallons or liters per minute (gpm or l/min).

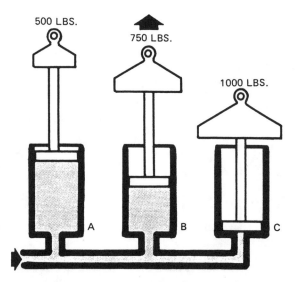

Figure 1.11. Pressure in a parallel connection.

Figure 1.12 illustrates the distinction between the two. In the diagram, liquid is pumped at the constant rate of 1 gpm through a chamber with two sections of different diameters. Section A is 2 ft long and Section B is 1 ft long, yet each section holds one gallon. Therefore, each section will be emptied and refilled every minute.

Since the sectional area of A is half that of B, the liquid must travel through Section A at 2 fpm. Liquid traveling through Section B moves only half as fast, or 1 fpm.

If A and B are sections of a hydraulic line, it can be seen that as the line increases in size, the velocity of the liquid will decrease if a constant rate of flow is maintained. On the other hand, decreasing the size of the line will increase the velocity. In fact, if a line in a hydraulic system were substituted with one whose diameter is doubled, the sectional area will be four times as large and the velocity of the oil will be ¼ its original velocity. If the line diameter is ½ the original size, the velocity will increase four times to maintain a constant rate of flow. Stated another way:

- The velocity of a fluid is inversely proportional to the sectional area of the line through which it passes.

Low velocity is usually desirable in a hydraulic system to reduce friction and turbulence, providing there is a sufficient rate of flow for proper system operation.

From this, it can be seen that both flow rate and velocity are closely related to the speed at which a load is moved. In Figure 1.13, the small-diameter cylinder (A) will move its piston faster than the large-diameter cylinder (B) if the rates of flow (gpm) are equal.

However, if the rate of flow to both cylinders is doubled, each cylinder will fill in half the time and the speed of each piston will consequently double. We can therefore conclude that:

- The speed of the piston of a cylinder is directly proportional to the flow.

Pressure Drop

Fluid flows because of a difference in pressure, and the pressure of a flowing fluid is always higher upstream and lower downstream. Therefore, the pressure differential of a flowing fluid is referred to as pressure drop. (Figure 1.14)

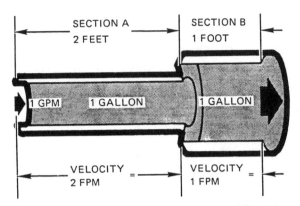

Figure 1.12.   Relationship between velocity and flow rate.

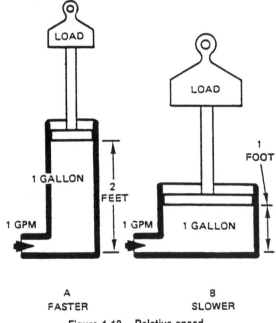

Figure 1.13.   Relative speed.

Fluids passing through a hose, a valve, or any restriction, no matter how much resistance to flow it offers, will produce some amount of pressure drop and heat. The greater the restriction, the greater the pressure drop and the higher the heat generated.

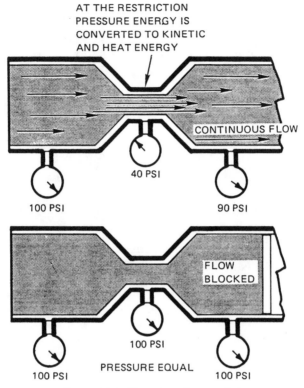

Figure 1.14. Flow through a restriction.

The principle derived from this phenomenon states that:

■ As the velocity of the fluid in a system increases, its pressure decreases.

Of course, as soon as the fluid is confined, pressure equalizes throughout the system.

Power

Power is the rate of doing work or the rate of energy transfer. To visualize power, think about climbing a flight of stairs. If you walk up, it is pretty easy. But, if you run up, you are liable to get to the top out of breath. The same amount of work was done either way . . . but when you ran up, you did it at a faster rate, which required more power.

The standard unit of power is the horsepower (hp). This was devised by James Watt to relate the ability of his steam engine to the pulling power of a horse. By experimenting with weights, pulleys, and horses, Watt decided that a horse could comfortably do 550 foot-pounds of work per second (ft-lb/sec), or 33,000 foot-pounds per minute (ft-lb/min), hour after hour. This value has since been designated as one horsepower. Power, then, is force multiplied by distance divided by time:

$$P \text{ (power)} = \frac{F \text{ (force)} \times D \text{ (distance)}}{T \text{ (time)}}$$

1 hp = 33,000 foot-pounds/minute (ft-lb/min)
     = 550 foot-pounds/second (ft-lb/sec)
     = 746 watts (w) (electrical power)
     = 42.4 British Thermal Units of Heat/minute(Btus/min)

The power used in a hydraulic system can be computed if the flow rate and the pressure are known:

$$hp = \frac{gpm \times psi}{1,714}$$

or

$$hp = gpm \times psi \times .000583$$

How a Hydraulic System Works

The most basic hydraulic system has two parts:

1. The *pump* that moves the oil
2. The *cylinder* that uses the moving oil to do work.

In Figure 1.15, when force is applied to the lever, the hand pump forces oil into the cylinder. The pressure of this oil pushes up on the piston and lifts the weight.

In effect, the pump converts a mechanical force to hydraulic power, while the cylinder converts the hydraulic power back to mechanical force to do work.

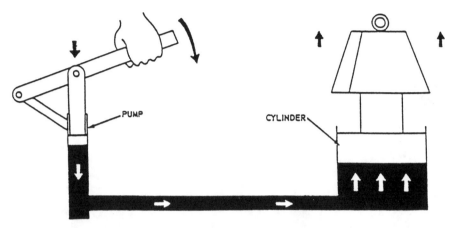

Figure 1.15.   Basic hydraulic system.

It should be noticed that the pump is smaller than the cylinder. This means that each stroke of the pump would only move enough oil to move the piston a small amount. However, the load lifted by the cylinder is much greater than the force applied to the pump piston. To lift the weight faster, the pump must be worked faster, increasing the volume of oil to the cylinder.

The system that was just described is a system that might be found on a hydraulic jack or a hydraulic press; however, to meet the hydraulic requirements in most other applications, a greater quantity of oil must be provided at a more consistent rate. Also, control of the oil movement is required, as shown in Figure 1.16.

Some additional components have been introduced in the circuits shown on Figures 1.17 and 1.18. These provide a means for regulating the maximum system pressure and the load direction. The lever-operated pump has been replaced with a gear-type pump. This is one of many types of pumps that transform the rotary force of a motor or engine into hydraulic energy.

The *control valve* directs the oil to allow the operator to control the constant supply of oil from the pump to and from the hydraulic cylinder. When the control valve is in the neutral position, as shown in Figure 1.17, the flow of oil from the pump goes directly through the valve to a line that carries the oil back to the reservoir. At the same time, the valve has trapped oil on both sides of the hydraulic cylinder, thus preventing its movement in either direction.

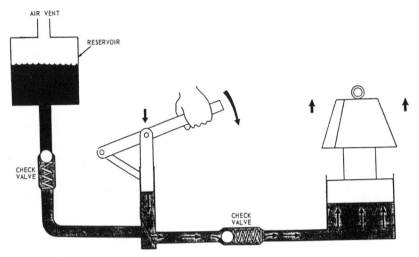

Figure 1.16. Hydraulic system with reservoir and check valves added.

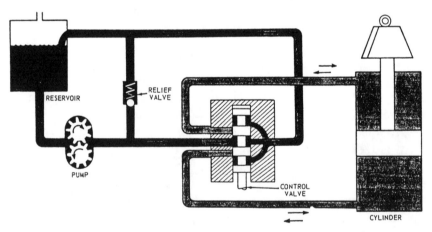

Figure 1.17. Hydraulic system with relief valve and double-acting cylinder added.

When the control valve is moved down (Figure 1.18), the pump oil is directed to the cavity on the bottom of the cylinder piston, pushing up on the piston and raising the weight. At the same time, the line at the top of the cylinder is connected to the return passage, thus allowing the oil forced from the top side of the piston to be returned to reservoir.

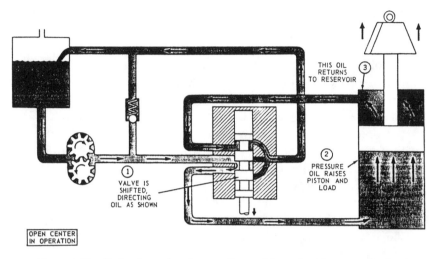

OPEN CENTER
IN OPERATION

Figure 1.18.   Hydraulic system in operation—raising a load (open-center type).

When the control valve is moved up, oil is directed to the top of the cylinder, lowering the piston and the weight. Oil from the bottom of the cylinder is returned to the reservoir.

The *relief valve* protects the system from high pressures. If the pressure required to lift the load is too high, this valve opens and relieves the pressure by returning the oil back to the reservoir. The relief valve is also required when the piston reaches the end of the stroke. At this time there is no other path for the oil and it must be returned to the reservoir through the relief valve.

The Pros and Cons of Hydraulics

As was seen in the simple hydraulic system just described, the purpose is to transmit power from a source (engine or motor) to the location where this power is required for work. In order to consider the advantages and disadvantages of the hydraulic system, one can compare it to the other common methods of transferring this power. These would be mechanical (shafts, gears, or cables) or electrical.

Advantages

1. **Flexibility.** Unlike the mechanical method of power transmission where the relative positions of the engine and work site must remain relatively constant with the flexibility of hydraulic lines, power can be moved to almost any location.

2. **Multiplication of force.** In the hydraulic system, very small forces can be used to move very large loads simply by changing cylinder sizes.

3. **Simplicity.** The hydraulic system has fewer moving parts, fewer points of wear, and it lubricates itself.

4. **Compactness.** Compare the size of a small hydraulic motor with an electric motor of equal horsepower. Then imagine the size of the gears and shafts that would be required to create the forces that can be attained in a small hydraulic press. The hydraulic system can handle more horsepower for its size than either of the other systems.

5. **Economy.** This is the natural result of the simplicity and compactness of the system, which translates to relatively low cost for the power transmitted. Also, power and frictional losses are comparatively small.

6. **Safety.** There are fewer moving parts such as gears, chains, belts, and electrical contacts than in other systems. Overloads can be more easily controlled by using relief valves than is possible with the overload devices on other systems.

Disadvantages

1. **Efficiency.** While the efficiency of the hydraulic system is much better than the electrical system, it is lower than for the mechanical transmission of power.

2. **Need for cleanliness.** Hydraulic systems can be damaged by rust, corrosion, dirt, heat, and breakdown of fluids. Cleanliness and proper maintenance are more critical in the hydraulic system than in the other methods of transmission.

Types of Hydraulic Systems

Two major types of hydraulic systems are used today: *open-center systems* and *closed-center systems.*

The simple hydraulic system described (Figure 1.17) is an open-center system. This system requires that the control valve spool be open in the center to allow pump flow to pass through the valve and return to the reservoir. The pump used supplies a constant flow of oil and

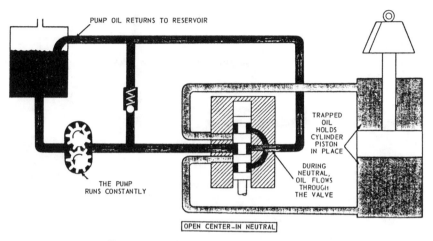

Figure 1.19. Open-center system in neutral.

the oil must have a path for return when it is not required to operate a function (see Figure 1.19).

In the closed-center system, the pump is capable of "taking a break" when oil is not required to operate a function. Therefore, the control valve is closed in the center, which stops (dead-ends) the flow of oil from the pump—the "closed center" feature (see Figure 1.20).

In neutral, the oil is pumped until pressure rises to a predetermined level. Then a pressure-regulating valve allows the pump to shut itself off and to maintain this pressure to the valve. This necessitates the use of a variable displacement pump, which will be described later.

When the control valve is operated as shown in Figure 1.21, oil is diverted from the pump to the bottom of the cylinder. The drop in pressure caused by connecting the pump pressure line to the bottom of the cylinder causes the pump to go back to work, pumping oil to the bottom of the piston and raising the load.

When the valve was moved, the top of the piston was connected to a return line, thus allowing return oil forced from the piston to be returned to the reservoir or pump.

When the valve is returned to neutral, oil is again trapped on both sides of the cylinder and the pressure passage from the pump

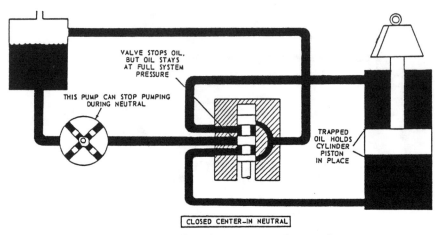

Figure 1.20. Closed-center system in neutral.

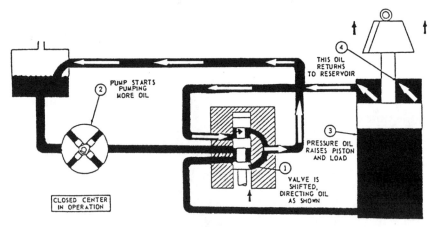

Figure 1.21. Closed-center system in operation—raising a load.

is dead-ended. At this time, the pump again takes a break—it stops pumping because the displacement becomes nearly zero—there is no flow from the pump.

Moving the spool in the downward position directs oil to the top of the piston, moving the load downward. The oil from the bottom of the piston is sent into the return line.

With the closed center system, if the load exceeds the predetermined standby pressure or if the piston reaches the end of its stroke, the pressure buildup simply tells the pump to take a break. The need for a relief valve is only to protect the system in the event the pump control malfunctions.

Hydraulic Facts

Some key facts that will help in the understanding of hydraulics are as follows:

1. Hydraulic power is nearly always generated from mechanical power. Example: A hydraulic pump driven by an engine crankshaft.
2. Hydraulic power output is nearly always achieved by converting back to mechanical energy. Example: A cylinder that raises a heavy load.
3. There are three types of hydraulic energy: potential or pressure energy; kinetic energy, the energy of moving liquids; and heat energy, the energy of resistance to flow, or friction.
4. Hydraulic energy is neither created nor destroyed, only converted to another form.
5. All energy put into a hydraulic system must come out either as work (gain) or heat (loss).
6. When a moving liquid is restricted, heat is created and there is a loss of potential energy (pressure) for doing work. Example: A tube or hose that is too small or is restricted. Orifices and relief valves are also restrictions, but they are purposely designed into systems.
7. Flow through an orifice or restriction causes a pressure drop.
8. Oil must be confined to create pressure for work. A tightly sealed system is a must in hydraulics.
9. Oil takes the course of least resistance.
10. Oil is normally pushed into a pump, not drawn into it. (Atmospheric pressure supplies this push. For this reason, an air vent is needed in the top of the reservoir.)
11. A pump does not pump pressure; it creates flow. Pressure is caused by resistance to flow.
12. Two hydraulic systems may produce the same power output— one at high pressure and low flow, the other at low pressure and high flow.
13. A basic hydraulic system must include four components: a reservoir to store the oil; a pump to push the oil through the system; valves to control oil pressure and flow; and a cylinder (or motor) to convert the fluid movement into work.
14. There are two major hydraulic systems:

    - Open-center system: Pressure is varied but flow is constant.
    - Closed-center system: flow is varied but pressure is constant.

15. There are two basic types of hydraulics:

- Hydrodynamics: The use of fluids at high speeds "on impact" to supply power. Example: a torque converter.
- Hydrostatics: The use of fluids at relatively low speeds but at high pressures to supply power. Example: most hydraulic systems.

Pneumatics

Introduction

Each time you take a breath, you are inhaling a little more than $^3/_{100}$ of a cubic foot of air—a colorless, odorless, tasteless mixture of gases without which one could not survive for more than 5 or 6 minutes. The air we breathe is the same air we compress and use in an endless variety of ways vital to our industrial complex. Our current knowledge of air is the result of centuries of supposition, experience and experiments, inspired guesswork, and hard labor by hundreds of scientists and philosophers.

Mining operations, in the past as well as now, had the problem of water seeping into the bottom of shafts and excavations. Simple hand pumps were devised to bring this water to the surface. These pumps consisted of a cylinder into which a piston was fitted. A handle was attached to the piston so that when the handle was pushed down, the piston was raised in the cylinder, creating a vacuum. Under the cylinder, a hollow shaft extended down into the water and, through an arrangement of check valves, the water rushed into the cylinder "to fill the vacuum" and was subsequently pushed out by the piston on the next stroke of the handle.

No matter how vigorously the handle was pumped or how well the piston was fitted to the cylinder, the water could not be raised higher than about 30 ft or 9.14 m.

This failure of suction was of interest to one of Galileo's students— Evangelista Torricelli. Torricelli suspected that the suction of the pump (the vacuum in the cylinder) was not responsible for pulling the water up, but rather that the water was pushed into the cylinder by the pressure or weight of the air once a partial vacuum was established by the pump. In other words, he suspected that the air had a finite weight equal to a 33-ft column of water.

To test this, he filled a glass tube about one yard long with mercury. One end of the tube was sealed. Holding his thumb over the open end of the tube, he upended the tube into a dish partially filled with mercury. On removing his thumb, the mercury flowed out of the tube into the dish and then stopped flowing when the column fell to a height of about 30 in. (76.2 cm). The weight of the air (atmosphere) exerted a force equal to 30 in. of mercury. Since mercury is 13.6 times as dense as water, this was equal to about 34 ft of water. Indeed, if a good enough vacuum could be established, 33.9 ft (10.33 m) is the highest the atmosphere would "lift" a column of water, which helped to explain the mine-and-pump problem.

Robert Boyle, an English physicist who lived in the latter part of the 17th century, established the relationship between the pressure and volume of air in another experiment using glass tubes and mercury.

He partially filled a "J"-shaped glass tube with mercury. He then plugged the short leg and proceeded to pour additional mercury into the long leg of the tube. He observed that, when he doubled the exerted pressure of the mercury in the long leg of the "J," the air space above the mercury in the short leg was reduced by half. In other words, as he doubled the pressure, he reduced the volume of the air by half. This is *Boyle's Law*, which states that "with a constant temperature, volume is inversely proportional to pressure."

About the same time that Boyle was developing his theories, the French physicist Blaise Pascal conducted a series of experiments involving the behavior of gases and liquids. As mentioned, he proved that pressure set up in a liquid or gas exerts a force equally in all directions.

From his investigation and experiments, Jacques-Alexandre-César Charles gave his name to the other principal law—along with Boyle—that makes up the ideal gas law. *Charles's Law* was officially enunciated in 1787. One way of stating this law is: "With a constant volume, absolute pressure is directly proportional to absolute temperature." That is to say, if you double the absolute temperature keeping a constant volume, you will double the pressure. Charles's Law is what is in mind when you are cautioned against throwing empty aerosol cans into a fire. As the remaining gas in the can is heated, the pressure might increase sufficiently to explode the container (Figure 1.22).

Absolute temperature is used when applying Charles's Law, to avoid the confusion that might occur with degrees, Fahrenheit, or Centrigrade. Absolute temperature starts at absolute zero, which is −460°F, or 0°R (zero degrees Rankine). Rankine was the scientist who discovered absolute zero. On this basis, 0°F equals 460°R, and 60°F equals 520°R (460°R + 60°F).

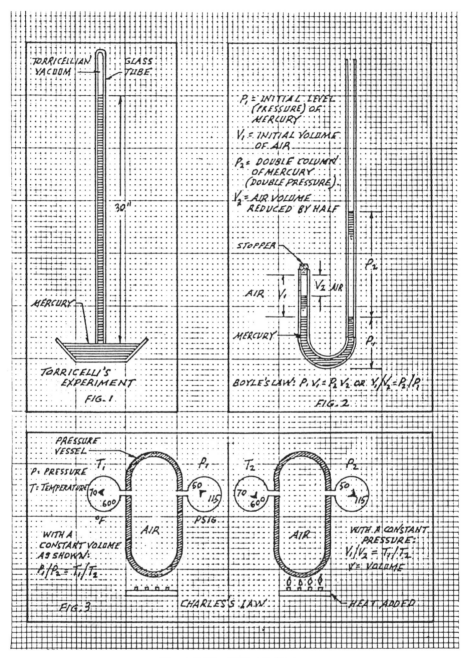

Figure 1.22. Basic gas laws.

The amount of water vapor in the air is described by the words *relative humidity*. The capacity for air to hold water in vapor or gaseous form is primarily a function of the temperature of the air, so any explanation of relative humidity must make specific reference to the temperature of the air. The higher the temperature, the greater the capacity to hold water vapor. Briefly, relative humidity is defined as the amount of water vapor in the air at a given temperature, relative to its total capacity at the same temperature. By withdrawing heat energy from air you can dry air—that is, condense a portion of the water vapor; in fact, heat exchangers are used extensively as devices for drying air.

When the relative humidity of air is 100%, the point where it can hold no additional water has been reached. This is also called the *dew point* of the air. The water vapor will begin to transform, or condense, into liquid "dew".

Standard Air and Free Air

In the United States, air is most commonly measured by the cubic foot (ft³). The dynamics of air are determined by the air's pressure, temperature, and water vapor content (relative humidity). Therefore, industry has adopted a standard on air specifying these three factors. Pressure differentials, pressure drop, flow, and performance criteria of various pneumatic equipment and components are based and designed around this "standard" air. A standard cubic foot (scf) of air is: one cubic foot volume at sea-level atmospheric pressure (14.69 lb/in²), 68°F temperature, and 36% relative humidity. This is the standard established by the ASME—the American Society of Mechanical Engineers. For air to perform some useful work, there must be pressure differentials. Flowing air (a function of pressure difference) is measured by the cubic foot volume past a given point during some period of time, usually by the minute; so we have scfm or standard cubic feet per minute.

It is well to make a distinction between the weight of air and the pressure of air. Weight is the air's density or mass. If you have a truck wheel and tire that weigh 150 lb empty, the wheel and tire may weigh 151 lb when inflated. Pressure is the force that air exerts against a surface, usually expressed as pounds per square inch (psi).

While it is well to know the definition of a scf of air, you will rarely encounter air under these precise conditions. Rather, you will be normally dealing with *free air*. Free or *actual* air is ambient air with its constantly changing temperature and relative humidity. Atmospheric pressure changes with altitude.

Air is "springy," and this compressibility of air means you can keep squeezing more and more air into a constant volume (or storage tank).

As more air is forced into this constant volume, the pressure rises. This ratio of free air required to achieve elevated pressure in a constant volume is expressed by the following equation:

$$\frac{\text{Atmospheric Pressure + Gauge Pressure*}}{\text{Atmospheric Pressure}} \quad \text{or} \quad \frac{\text{psia}}{\text{atmos.}}$$

Therefore, to determine how many cubic feet of free air must be pumped into a cubic foot container to get 100-psi gauge reading, at sea level, this ratio is calculated as follows:

$$\frac{114.69}{14.69} = 7.8, \text{ a ratio of 1 to 7.8}$$

and at a 5,000 ft elevation the ratio is:

$$\frac{112.22}{12.22} = 9.18, \text{ a ratio of 1 to 9.18}$$

It can be seen that, at sea level, it requires 7.8 ft³ of free air to make 1 ft³ of air at 100 psig (Figure 1.23), while at 5,000 ft elevation, it would require 9.18 ft.³

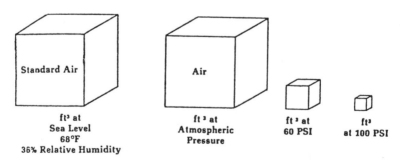

Figure 1.23. Standard, free, and compressed air.

* *Pressure gauges normally show only elevated pressures. Gauge pressure is psig—pounds per square inch gauge. The sum of gauge pressure and atmospheric pressure is called absolute. Knowing and using the absolute pressure is also required in calculating flow and pressure drop on various components.*

Pneumatic Principles

The term "pneumatic system" can mean many things in an industrial plant. If the plant does heavy fabricating, pneumatics probably include air tools and some hoisting equipment. If the plant produces small packaged materials, pneumatics can mean air-operated machinery, cylinders, and clamping devices. If the plant processes raw materials, pneumatics can mean air cylinders, air motors, process controls, pneumatic conveyors, and possibly fluidic devices that control the sequence of equipment operation.

Force, Weight, and Mass

A force is a push or a pull exerted on an object to change its position or movement. This includes starting, stopping, and changing its speed or direction of movement. In a pneumatic system, force must be present at all times for the system to function. This force is compressed air.

An object or substance has weight as a result of the gravitational force or pull on the object. In a pneumatic system, the compressor, tank, lines, any of the components, and even the air in the system have weight. This is true whether the air is held in the tank or is moving through the system.

All objects or substances have mass. The mass represents the amount of matter in an object and its inertia, or resistance to movement. The *mass* of an object determines its weight on earth or in any other gravitational field. The *inertia* of an object determines how much force is required to lift or move an object or to change its speed or direction of movement.

The *density* of an object is its weight for a specific volume or unit of measure. The density of a ft³ of "bone-dry" air at atmospheric pressure and a temperature of 60°F is 0.076 lb. Air is very light compared to water or hydraulic fluid. Its relatively low density makes it suitable for long-distance and high-speed control applications. Also, because of its low weight and inertia, it will not cause pneumatic shocks as hydraulic fluid does when a valve is closed quickly.

Work and Energy

Work takes place when a force (in lb or kg) moves through a distance (in in., ft, or m). The amount of work done is expressed in the English system of measurement in ft-lb or in.-lb, as shown in the following formula:

$$\text{Work} = \text{Force (lb)} \times \text{Distance (ft or in.)}$$
$$= \text{ft-lb or in.-lb}$$

In a pneumatic system, the force in lb is exerted by air pressure acting on the area of a moving piston in a cylinder. As the piston moves, the pneumatic force acts through the length of the stroke.

Power is defined as the amount of work (ft-lb) done in a given length of time (seconds or minutes), or ft-lb/min.

For the amount of power calculated to be meaningful, it must be compared with a unit of measurement. The common unit of power measurement is horsepower.

To use power and do work, energy must be expended. The Law of Conservation of Energy states that "Energy cannot be created or destroyed. It can only be transformed." Therefore, you will usually start off by using one kind of energy to get other kinds of energy. Some of this energy does useful work. Some of it only overcomes friction. The energy that overcomes friction is not lost, but is changed into heat energy.

The types of energy used in pneumatic systems include the following:

- **Electrical energy.** Operates the compressor motor.
- **Pneumatic energy.** Produced by the compressor.
- **Kinetic energy.** Produced when the compressed air is lifting an object.
- **Potential energy.** What the lifted object now has.
- **Heat energy.** Produced by friction in the compressor motor, the compressor, the moving air, and the moving piston.

Diffusion and Dispersion

Diffusion can be described as the rapid intermingling of the molecules of one gas with another. This should not be confused with evaporation, which is the changing of a liquid to a gas. To prevent compressed gases from rapidly diffusing into the surrounding air, they must be stored in closed containers.

Dispersion can be described as the temporary mixing of liquid particles with a gas. When air is compressed, compressor lubricating oil is picked up by the moving air in the compressor and dispersed in fine particles that remain suspended in the air for a time. If enough heat is generated in the compressor, some of the oil evaporates and is diffused in the air.

Although diffusion and dispersion should be kept to a minimum, liquids (especially water) evaporate into the surrounding gases constantly. Because liquids are heavier than gas, they do not mix readily. Therefore, when air and suspended water or oil are put in the same tank or flow through the air lines, the water or oil will settle out and flow to the lowest places. That is why water usually collects at the bottom of a vertical air line, where it should be removed through a drain valve.

Air Flow in Pipes

Streamline or laminar flow is the ideal type of air flow in a pneumatic system because the air layers move in nearly parallel lines. Like all fluids, the layer of air next to the surface of the pipe moves the most slowly because of the friction between the fluid and the pipe. The layer of moving fluid (air) next to the outermost layer moves a little faster, and so on, until the fluid layers nearest the center of the flow passage move the most quickly.

Turbulent flow conditions usually occur because the flow passage is too small for the desired flow velocity of the air. The density and viscosity of the air also affect turbulent flow, but not as much as the flow passage and the flow velocity.

Rough or irregularly formed air passages, sudden enlargement or reduction in the diameter of the flow passages, and sudden changes in the direction of flow should all be avoided. When air must pass through a passage of reduced size, the restriction should be smooth and gradual.

Turbulent flow heats the air, wastes power by requiring higher air pressure, and can damage the flow passages and ports in the pneumatic equipment.

Pneumatic Power Systems

All industrial plants use a fluid power system of one type or another. Work is performed by a fluid under pressure in the system. The system may function as part of a process (heating or cooling), or it may be a secondary service system (compressed air). A fluid can be either a gas, such as compressed air, nitrogen, or carbon dioxide, or a liquid, such as oil or water. The word *pneumatic* is derived from the Greek work for unseen gas. Originally pneumatic referred only to the flow of air. Now it includes the flow of any gas in a system under pressure.

Some of the ways that pneumatic systems perform work include operating pneumatic tools, linear motion devices, door openers, and rotary motion devices. Pneumatic systems are also used to control flow valves in chemical process equipment and in large air-conditioning systems. In more sophisticated systems, pneumatics are used to operate sequencing control valves in much the same way as electrical relays.

The pneumatic system consists of a number of elements that enable it to perform its intended function. These system elements are shown on Figure 1.24.

First, an *air compressor,* driven by an electric or internal combustion engine, is needed to supply a source of pressurized air. A *pressure control* is used to limit the maximum air pressure. A *filter* screens out the dirt and grit from the air to protect the sensitive control components. A *regulator* provides a means for adjusting the pressure to any desired value. The *lubricator* adds a clean mist of oil into the air streams to reduce friction and wear. The *directional control* and *functional control* provide a means for controlling the speed and movement of the *actuator.*

The air compressor system is made up of a number of components—not just a compressor. A typical system consists of a compressor, drive motor, heat exchanger, and storage tank. The piping is also part of the system.

Compressor

Probably the most common type of compressor in use today is the reciprocating type. These employ a crankshaft that is driven by a motor or engine. Most reciprocating compressors bring the air to the final desired pressure through two or more "stages." This means that the free air is initially compressed to about 30 psig by one set of pistons. This partially pressurized air then flows into another set of pistons, which raises the pressure higher and so forth. Most common are two stages, which economically discharge the final air at typical supply pressure of 90 to as high as 150 psig.

Compressors that are found increasingly are "rotary" types that use screw-shaped rotating devices as the compressing elements. As these elements rotate in their housing, free air is drawn in at one end and is compressed as it is forced toward the discharge side.

The *volumetric efficiency* of a compressor is the ratio of the actual air delivered to the total displacement. This efficiency for the compressors

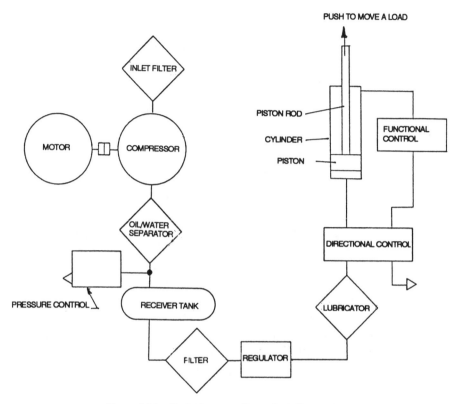

Figure 1.24. Typical pneumatic system diagram.

under consideration, converted to an easy to use figure, is from 4 to 4.2 scfm/hp. In other words, these electric motor-driven compressors will draw free air at the intake, pressurize it to a nominal 100 psig, and discharge it at the above rate. A 50-hp compressor, for example, has a 200-scfm capacity. As a matter of fact, one can use the 4-scfm/hp figure with reasonable assurance that the calculations will be sufficiently accurate for all practical problems and applications.

A common type of heat exchanger used to cool the compressor discharge air is the type that uses water as the cooling medium. These are water-to-air heat exchangers. Smaller tank-mounted compressors use air-to-air heat exchangers almost exclusively, but air-to-air exchangers are available for large compressors as well.

The size of the storage tank is usually commensurate with the compressor capacity and will range from 60 gal up to 600 gal and larger. Tank capacities are usually specified in liquid gal rather than ft³ (to convert from gal to ft³, multiply the gal by 0.1337).

Storage tanks perform two basic functions in the air system: they accumulate energy and dampen severe pulsations. Storage tanks are constructed in accordance with strict safety codes and are rated to safely withstand the system pressure. As a further safeguard against a catastrophic explosion, these tanks are fitted with safety relief valves.

Pneumatic Valves

Working energy that is transmitted through a pneumatic system must be directed and under complete control at all times. If not under control, useful work may not be done and machinery or machine operators might be harmed. One of the advantages of transmitting energy pneumatically is that energy can be controlled relatively easily.

The basic pneumatic valve is a mechanical device consisting of a body and a moving part that connects and disconnects passages within the body. The flow passages in pneumatic valves carry air. The action of the moving part may control system pressure, direction of flow, and rate of flow.

Pressure in a pneumatic system must be controlled at two points: at the compressor and after the air-receiver tank. Control of pressure is required at the compressor as a safety for the system. Control pressure at the point of air usage is necessary for safety and so that actuators receive the proper pressure to avoid wasting energy.

A compressor usually delivers compressed air to a receiver tank until high pressure is reached, then its air delivery is controlled. When air pressure in the tank decreases, the compressor delivery is increased (within its design limits). Compressor operation of this type is a power savings for the system. One way to accomplish this function is through the use of a pressure switch.

In a pressure switch, the system pressure is sensed at the bottom of the piston through the pressure-switch inlet. When pressure in the system is at its low level, the spring pushes the piston down. In this position, a contact is made, causing an electrical signal to turn on the compressor.

As pressure in the receiver tank rises, it forces the piston upward. With system pressure at its high level, the piston breaks the electrical contact, shutting down the compressor. However, it should be noted that in large compressors, the compressor is unloaded in some way to prevent the compression stroke from compressing any air.

The safety relief valve is a normally nonpassing (closed) valve. The poppet of the safety relief valve is seated on the valve inlet. A spring holds the poppet firmly on its seat. Air cannot pass through the valve until the force of the spring biasing the poppet is overcome. Air pressure at the compressor outlet is sensed directly on the bottom of the poppet. When air pressure is at an undesirably high level, the force of the air on the poppet is greater than the spring force. When this happens, the spring will be compressed and the poppet will move off its seat, and air will exhaust through the valve ports.

Good maintenance and safety procedures require the periodic checking of these safety valves to verify that the valve can move freely. Many times, this can be done by moving the manual test lever to open the valve.

The pressure regulator is a normally passing (open) valve. "Open" means that the flow passage normally allows air to flow freely. With a regulator positioned after a receiver tank, air from the receiver can expand (flow) through the valve to a point downstream of the secondary passage. When pressure in this passage of the regulator rises, it is transmitted through an internal pilot passage that leads to the piston area on the opposite side of the spring. The piston has a relatively large surface area exposed to secondary pressure and is therefore responsive to secondary pressure fluctuations. When the controlled pressure nears the preset level, the piston moves upward allowing the poppet to move towards its seat to control (or throttle) the flow. The poppet blocks flow once it seats and does not allow pressure to continue building downstream. In this way, air at a controlled pressure is made available to an actuator downstream. Also, a pressure regulator ensures that energy in an air system is not wasted. For instance, if an air receiver charged to 100 psia were allowed to subject its full pressure to an actuator that only required 60 psia, pressure energy would be wasted.

The double-acting cylinder has a port at each end of the cylinder body by which air under pressure can enter or exhaust. This causes the piston rod to extend or retract (double acting). In order to change the direction of air flow to and from the cylinder, a directional control valve is used.

The typical directional control valve consists of a valve body with four internal flow passages and moving part, a spool, which alternately connects a cylinder port to the supply pressure or to exhaust.

In a pneumatic system, actuator speed is determined by how quickly the actuator can be filled and exhausted of air. In other words, the speed of a pneumatic actuator depends on the force available from the pressures acting on both sides of the piston, a result of the ft³/min of air flowing into the inlet and out of the exhaust port.

A pressure regulator influences actuator speed by portioning out to the circuit the pressure required to equal the load resistance at the actuator. This additional pressure is used to develop air flow. Even though this is the case, pressure regulators are not used to vary actuator speed. In a pneumatic system, actuator speed is affected by a restriction, often called a "flow-control" valve. This valve will meter flow at a constant rate only if the system resistance is constant and the pressure at the inlet of the cylinder and its outlet do not vary throughout the entire work cycle.

A needle valve in a pneumatic system affects the operation by causing a restriction in order to control actuator speed. The typical needle valve consists of a valve body and an adjustable part that can be a tapered-nose rod that is threaded into the valve body. The more the tapered-nose rod is screwed towards its seat in the valve body, the greater the restriction to free flow.

By restricting exhaust air flow in this manner, a back pressure is generated within the actuator, thus reducing the forces available to create motion. This means a larger portion of regulator pressure is used to overcome the resistances at the actuator and less pressure energy is available to develop flow.

With less ft³/min of air flowing into the actuator, the actuator speed decreases.

Hydraulic System Components

Hydraulic Pumps

The pump is the heart of the hydraulic system. It creates the flow of fluid that supplies the whole circuit. The human heart is a pump. So was the old water pump used on the farm. Somewhere in between, engineers have devised many kinds of hydraulic pumps, which do more than the old water pump, but only strive for the perfection of the human heart pump.

A hydraulic pump is one that moves fluid and induces fluid to do work—in other words, *a pump that converts mechanical force into hydraulic fluid power.*

"Displacement" is the volume of oil moved or displaced during each cycle of the pump. That is, the amount of oil coming out of the pump during each revolution that it makes.

In this sense, hydraulic pumps fall into two broad types: *fixed displacement pumps* and *variable displacement pumps.*

Fixed displacement pumps move the same volume of oil with every cycle. This volume is only changed when the speed of the pump is changed. Volume can be affected by the pressure in the system, but this is due to an increase in leakage back to the pump inlet. Usually this occurs when pressure rises.

Variable displacement pumps can change or vary the oil they move with each cycle, even at the same speed. These pumps have an internal mechanism that varies the output of oil, usually to maintain a constant pressure in the system. As pressure rises, the volume decreases. Remember, a hydraulic pump does not create pressure; it creates flow. Pressure is caused by *resistance* to flow.

The essentials of any hydraulic pump are:

- An inlet port that is supplied fluid from a reservoir,
- An outlet port connected to the pressure line,

- Pumping chambers to carry the fluid from the inlet to the outlet port, and
- A mechanical means for activating the pumping chambers—a motor or engine to drive the pump.

In most hydraulic pumps, the design is such that the pumping chambers increase in size at the inlet, thereby creating a partial vacuum. The chambers then decrease in size at the outlet to push the fluid into the system. The vacuum at the inlet is used to create a pressure difference so that fluid will flow from the reservoir to the pump. In some systems, the inlet is charged or supercharged; that is, a positive pressure rather than a vacuum is created by a pressurized reservoir, a head of fluid above the inlet, or even a low-pressure charging pump.

There are many different types of pumps used in hydraulic systems. The three basic designs used are: *gear pumps, vane pumps,* and *piston pumps.*

We will show how each type of pump operates and how it is used. A hydraulic system may use one of these pumps, or it may use two or more in combination.

Gear Pumps

Gear pumps are the "workhorses" of hydraulic systems. They are widely used because they are simple and economical. While not capable of a variable displacement, they can produce the volume needed by most systems using fixed displacement. Often they are used as charging pumps for larger system pumps of other types.

Two basic types of gear pumps are used: *external gear pumps* and *internal gear pumps.*

External Gear Pump

The external gear pump (Figure 2.1) consists of two gears enclosed in a closely fitted housing. The gears rotate in opposite directions and mesh at a point in the housing between the outlet and inlet ports. This type of pump may use spur, helical, or herringbone gear-tooth forms.

As the gears rotate, a partial vacuum is formed, which causes liquid to be forced in through the inlet port. The liquid is then trapped in the space between the teeth of the two revolving gears and the housing. Fluid from the discharge side cannot return to the intake side because of the close meshing of the two gears and the small clearances

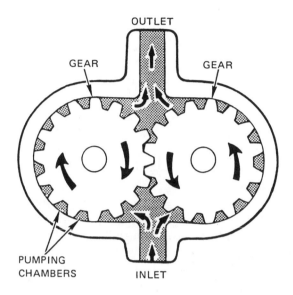

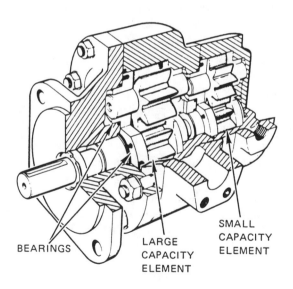

Figure 2.1. External gear pump.

between the gears and housing. The close meshing of the gear teeth provides a seal between the inlet and outlet ports. The fluid accumulates on the outlet side and is forced out of the pump into the hydraulic system.

Internal Gear Pump

The internal gear pump (Figure 2.2) has the same characteristics as the external gear pump just described. The pump consists of two gears; one is an external gear, the other an internal gear. The internal gear gets its name from the fact that the gear teeth point in toward the axis instead of away from it. The external gear is mounted on a shaft and is positioned within the internal gear. As the pump operates, the inner gear drives the outer gear; each gear rotates in the same direction. The teeth of both gears engage only at one point in the pump. As the teeth separate, pockets are formed by the gear teeth; it is the change in volume of these pockets that creates a pumping action. Immediately after the meshing of the teeth, the volume of the pockets increases, causing fluid to be drawn through the inlet port. Thus, the fluid is trapped in the pockets. As the gears rotate, the volume of the pockets decreases and the fluid is forced out of the pump. The inlet side is in the area where the pockets increase and the discharge side is in the area where the pockets decrease.

A *Gerotor pump* is an internal gear pump with an inner gear and an outer gear. The inner gear has one less tooth than the outer gear. As the inner gear is turned by a prime mover, it rotates the larger outer gear. On one side of the pumping mechanism, an increasing volume is formed as gear teeth unmesh. On the other half of the pump, a decreasing volume is formed. A Gerotor pump has an unbalanced design.

The output volume of a gear pump is determined by the volume of fluid each gear displaces, and by the rpm at which the gears are turning. Consequently, the output volume of gear pumps can be altered by replacing the original gears with gears of different dimensions or by varying the rpm.

Gear pumps, whether of the internal or external variety, do not lend themselves to a change in displacement while they are operating. There is nothing that can be done to vary the physical dimensions of a gear while it is turning.

Vane Pumps

A vane pump (Figure 2.3) makes use of centrifugal force and it makes use of variable volume chambers. Centrifugal force moves the pump vanes into position to form the chambers. The expanding and diminishing of these chambers develops the flow of the fluid in and out of the pump.

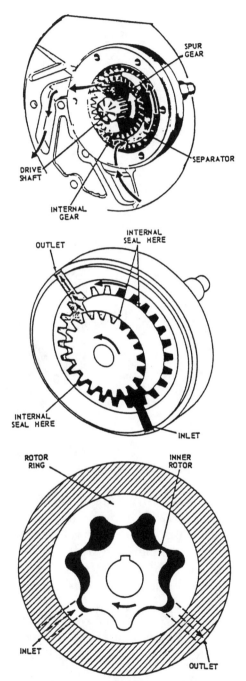

Figure 2.2.   Internal gear pump.

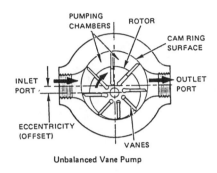

Unbalanced Vane Pump

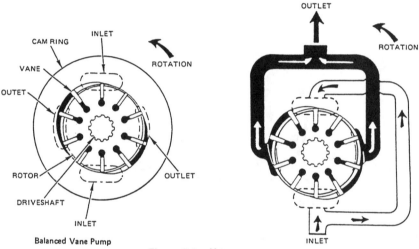

Balanced Vane Pump

Figure 2.3. Vane pumps.

Vane pumps are used extensively in industry to provide hydraulic power. Wear does not appreciably decrease efficiency, because the vanes always maintain close contact with the cam ring with which they rotate.

The principal parts of the pump are a slotted rotor, vanes, a cam ring with an elliptical inner contour, end caps, and bearings. The rotor is slotted and is driven by a shaft. The slots are cut at a slight angle to the outer circumference of the rotor. They are slanted toward the direction of rotation. Each slot of the rotor serves the purpose of holding a flat, rectangular vane. The vanes are free to move radially in the slot. Specifically, they can move in and out from the axis of rotation. However, they are prevented from moving sideways by end caps that retain the pump assembly.

The vanes are flat rectangular pieces of material. Each vane has three squared edges and one beveled edge. The beveled edge rides along the elliptical-contoured surface of the cam ring casing with the beveled edge trailing. As the rotor turns, centrifugal force ejects the vanes outward to contact and follow the elliptical cam ring. By this action, the vanes divide the area between the rotor and the ring into a series of chambers. These chambers vary in volume according to their respective positions about the rotor.

The ring that encloses the rotor and the vanes is circular in its outer contour, whereas the inner contour is oval or elliptical. The ring is anchored securely in the housing of the pump; it does not move. Since wear must be minimized, the cam-shaped ring is made of hardened and ground steel. It is the ring contour that makes it possible for each chamber to change in volume as it moves around the ring. Because of the ring contour, each chamber formed by the vanes changes in volume as it moves around the ring. Fluid is drawn into the pump as each chamber increases in volume, and as each chamber decreases in volume the fluid is ejected from the pump. Inlet and outlet ports are located at points to accommodate the intake and discharge areas of the pump.

Variable-Volume Vane Pumps

The amount of fluid that a vane pump displaces is determined by the difference between the maximum and the minimum distances the vanes are extended and the width of the vanes. While the pump is operating, nothing can be done to change the width of a vane. But, a vane pump can be designed so that the distance the vanes are extended can be changed. This is known as a *variable-volume vane pump* (Figure 2.4).

Variable-volume vane pumps are unbalanced pumps. Their rings are circular and not cam shaped. However, they are still referred to as cam rings.

The pumping mechanism of a variable-volume vane pump basically consists of a rotor, vanes, a cam ring that is free to move, a port plate, a thrust bearing to guide the cam ring, and something to vary the position of the cam ring. When the rotor is turned, an increasing and decreasing volume is generated and pumping occurs.

With the screw adjustment turned out slightly, the cam ring is not as off center to the rotor as before. An increasing and decreasing volume is still being generated, but not as much flow is being delivered by the pump. The exposed length of the vanes at full extension has decreased.

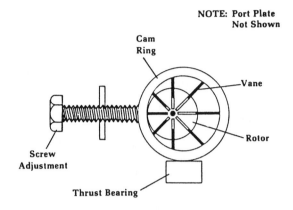

Variable Volume Vane Pump

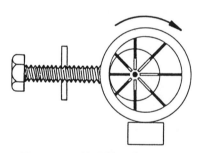

Figure 2.4.   Variable-volume vane pump.

With the screw adjustment backed completely out, the cam ring naturally centers with the rotor. No increasing and decreasing volume is generated. No pumping occurs. With this arrangement, a vane pump can change its output flow anywhere from full flow to zero flow by means of the screw adjustment.

Generally, variable-volume vane pumps are also pressure compensated. A pump that is pressure compensated stops pumping at a preset pressure level. A pressure-compensated vane pump consists of the same parts as a variable-volume vane pump. But in addition, an adjustable spring is used to offset the cam ring. When the pressure acting on the inner contour of the cam ring is high enough to overcome the force of the spring, the ring centers and, except for leakage, pumping ceases.

System pressure is therefore limited by the setting of the compensator spring. This takes the place of a system's relief valve in effect.

Piston Pumps

Piston pumps generate a pumping action by causing pistons to reciprocate within a piston bore.

There are two main types of piston pumps: radial and axial types. The radial type is one in which the pistons and cylinders radiate from the axis of a circular cylinder block like the spokes of a wheel. The axial type is one in which the pistons and cylinders are parallel to each other and parallel to the axis of rotation.

Radial-Piston Pump

A piston moving back and forth in a cylinder can draw liquid in and then push it out. One piston in a pump would not be practical; however, a number of pistons can be put in the same cylinder block. The radial-piston pump (Figure 2.5) consists of a number of pistons arranged around a hub in a steel block, which is bored with accurately machined cylindrical compartments to accommodate the pistons. All cylinder holes are bored an equal distance apart and they connect with a hole that is bored in the center of the block.

The pump consists of a pintle, which remains stationary and is actually a valve; a cylinder block that revolves around the pintle and contains the cylinders in which the pistons operate; a rotor of hardened steel against which the piston heads press; and a slide block, which houses and supports the rotor. The slide block does not revolve, but the rotor revolves due to the friction set up by the sliding action between the piston heads and the inner contour of the rotor. The slide block, which in effect is a casing or housing, contains the rotor, which is free to revolve.

The center point of the rotor is different from the center point of the cylinder block. It is that difference that produces the pumping action. If the rotor has the same center point as that of the cylinder block, there will be no pumping action since the piston does not move back and forth in the cylinder as it rotates with the block.

These pumps are generally used in high-pressure and high-flow applications, such as forging presses. They are not generally used on machines such as dozers or cranes.

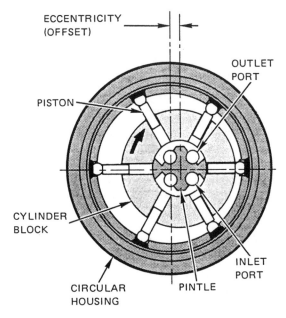

Figure 2.5.   Radial-piston pump.

Axial-Piston Pumps

In axial-piston pumps (Figure 2.6), the pistons stroke axially, or in the same direction as the cylinder block centerline. Axial-piston pumps may be the in-line design or the angle design.

In one type of in-line piston pump, the shaft and cylinder block are on the same centerline. Reciprocation of the pistons is caused by a swash plate that the pistons can run against as the cylinder block rotates. The drive shaft turns the cylinder block, which carries the pistons around the shaft. The piston shoes slide against the swash plate and are held against it by the shoe plate. The angle of the swash plate causes the pistons to reciprocate in their bores. At the point where a piston begins to retract, the opening in the end of the bore slides over the inlet slot in the valve plate and oil enters the bore through somewhat less than half a revolution. There is a solid area in the valve plate as the piston becomes fully retracted. As the piston begins to extend, the opening in the cylinder barrel moves over the outlet slot and oil is forced out the pressure port.

The displacement of the pump depends on the bore and stroke of the piston, and the number of pistons. The swash-plate angle determines

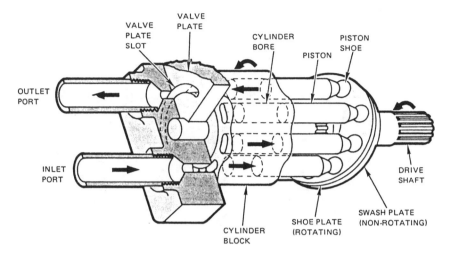

Figure 2.6. Axial-piston pump.

the stroke, which can be varied by changing the angle. In the fixed-angle unit, the swash plate is stationary in the housing. In the variable unit (Figure 2.7), it is mounted on a yoke, which can turn on pintles. Various controls can be attached to the pintles to vary pump delivery from zero to maximum. With certain controls, the direction of flow can be reversed by swinging the yoke past center. In the center position, of course, the swash plate is perpendicular to the cylinders and there is no piston reciprocation; therefore no oil is pumped.

Pressure Compensator Operation

The pressure-compensator control positions the yoke of a piston pump automatically to limit output pressure (Figure 2.8). It consists of a valve balanced between a spring and system pressure, and a spring-loaded yoke-actuating piston controlled by the valve.

The yoke return spring initially holds the yoke to a full delivery position. Pressure from the discharge of the pump is continually applied through passage "A" to the end of the compensator valve. The adjustment spring acting on the opposite end of the compensator valve opposes the fluid pressure in passage "A." When pressure in passage "A" is sufficient to overcome the load of the adjustment spring, the compensator valve moves, allowing fluid to enter the yoke-actuating piston. The fluid then forces the yoke-actuating piston to move the

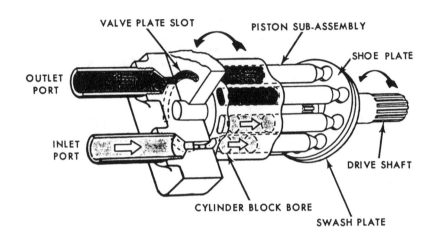

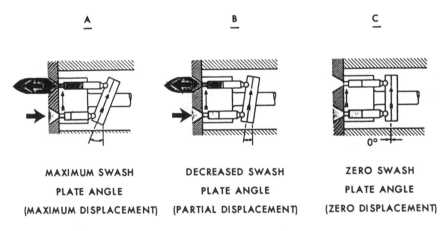

Figure 2.7. Variable-displacement piston pump.

yoke and decrease the stroke of the pistons in the cylinder block. Thus the delivery of the pump is reduced. If the pressure in passage "A" (outlet pressure) decreases, the adjustment spring will move the compensator spool, closing off passage "A" and permitting fluid in the yoke-actuating piston to drain through passage "B" to the housing. The yoke return spring then moves the yoke back toward its maximum delivery position.

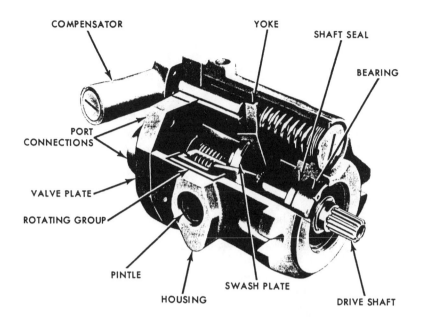

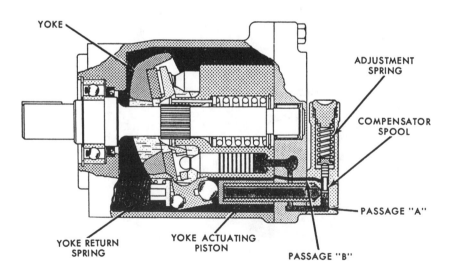

Figure 2.8. **Pressure-compensated axial-piston pump.**

The pump-compensator control thus reduces pump output to only the volume required to maintain a preset maximum pressure. Maximum delivery occurs when pressure is less than the compensator setting.

Bent-Axis–(Angle-) Type Pump

In the angle- or bent-axis–type piston pump (Figure 2.9), the piston rods are attached by ball joints to the drive shaft flange. A universal link keys the cylinder block to the shaft so that they rotate together but at an offset angle. The cylinder barrel turns against a slotted valve plate to which the ports connect. Otherwise pumping action is the same as the in-line pump.

The angle of offset determines the displacement of this pump, just as swash-plate angle determines an in-line pump's displacement. In fixed delivery pumps, the angle is constant. In variable models, a yoke mounted on pintles swings the cylinder block to vary displacement. Flow direction can also be reversed with appropriate controls.

Pump Delivery, Pressure, and Speed

Most pumps are rated by volume, which is usually expressed in gallons per minute (gpm). This rating is called several names—delivery rate, discharge, capacity, or size. Regardless of the rating, it cannot stand alone. It must be accompanied by a figure stating the amount of back pressure that the pump can withstand and still produce the gpm rating; because as pressure increases, internal pump leakage increases and usable volume decreases.

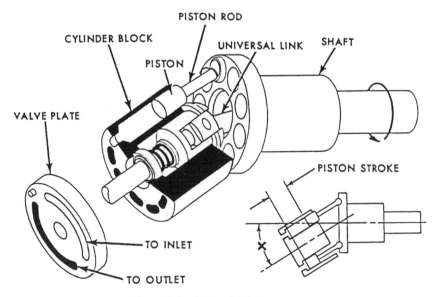

Figure 2.9. Bent-axis piston pump.

Pump speed must also be included with the volume rating for two reasons:

- In a fixed displacement pump, flow is directly related to the speed of the pump—the faster the speed, the more fluid is pumped.
- How fast the pump must go to produce a certain amount of flow indicates at what speed the driving mechanism for the pump must travel (in rpm). Add this to the delivery rate of a pump, and here is how a typical rating could read: "11.5 gpm with 2,000 psi at 2,100 rpm."

Pump quality is judged by three ratings: *volumetric efficiency, mechanical efficiency, overall efficiency.*

Volumetric efficiency is the ratio of the actual output of the pump to theoretical output (the amount it should put out under ideal conditions). The difference is usually due to internal leakage in the pump.

Mechanical efficiency is the ratio of the overall efficiency of the pump to volumetric efficiency. This difference is usually due to wear and friction on the pump's working parts.

Overall efficiency is the ratio of hydraulic power output to the mechanical power input of the pump. This is the product of both mechanical efficiency and volumetric efficiency.

Malfunctions of Pumps

The majority of pump failures are due to human errors, such as exceeding operating limits, improper installation, and the greatest cause, the use of oil that is dirty or of poor quality. Following are some of the causes of pump failures, and tips on how they can be prevented:

1. Contaminated fluid. As stated, contaminated or improper fluid is the biggest offender in hydraulic pump failures. Dirty fluid can damage a pump in many ways. Solid particles of dirt, sand, chips, etc. in the oil act as an abrasive on the pump's closely fitted parts. This causes abnormal wear on the parts and increases internal leakage, thus lowering the pump's efficiency. The reading obtained during a performance test of a system would be out of specification and lead to a change in the pump on the machine.

Sludge is formed by the chemical reaction of fluid due to excessive temperature change or condensation. It will build up on the pump's internal parts and eventually plug the pump. If the pump is plugged on the inlet side, it will be starved of oil and the resulting heat and friction will cause the pump parts to seize.

Air, plus water, plus heat will also create sludge by oxidation, and the results to the pump will be the same.

Water or other liquids in the fluid can rust the pump parts and housings. Rust will not only build up on the metal, but will also flake off into the fluid as abrasive solid particles.

2. Improper fluid. The selection of the proper fluid viscosity is extremely important. Viscosity is the rating given to the degree that the fluid resists flow. A high viscosity number indicates that the fluid is "heavy" and will strongly resist flow. A lower viscosity means a "lighter" fluid, which flows faster or easier.

Following are some of the things that can occur if the fluid viscosity is not correct:

- *Fluid is too "light."* Both internal and external leakage will increase. Pump slippage (efficiency) will be affected, causing heat and reduced efficiency. Parts wear will increase because of lack of adequate lubrication. System pressure will be reduced. Controls for the system will be "spongy".
- *Fluid is too "heavy."* Internal friction will increase, which, in turn, will increase flow resistance through the system. Temperature will increase, which can cause sludge build-up. Pressure drop throughout the system will increase. More power will be required for operation.

3. Overspeeding the pump. Increasing the pump speed will reduce the bearing life. The pump is rated to be used at a certain maximum speed. Its bearing life is based on that rated speed. By doubling the speed, the pump life will be reduced by one-half. Also, excessive speed can lead to pump cavitation, which is very destructive to the internal parts of the pump.

4. Cavitation. Cavitation occurs when the fluid does not entirely fill the space provided in the pump. This leaves air or vapor cavities in the liquid, which can be harmful to the pump.

The combination of the high velocity (speed) of the discharged fluid and a restriction, usually caused by a clogged inlet line, between the reservoir and the pump inlet causes the pressure of the incoming fluid to drop. When it is lowered, it cannot force enough fluid in to meet the demands of the pump. The result is that cavities or spaces are formed in the incoming fluid.

The pressure drops to the vapor tension of the fluid and the cavities fill with vaporized oil. The vapor tension of the fluid is that pressure at which, at a given temperature, the fluid boils and freely evaporates. This evaporation fills the cavities.

The problem is further complicated by the pressure drop, because this tends to release any dissolved air in the fluid and it, too, fills the cavities.

The damage to the pump results when these vapor-filled cavities, which have been formed in a low-pressure area, meet with a high-pressure area in the pump and are forced to collapse. This creates an action similar to an implosion, which disintegrates or chips away small particles of the metal parts of the pump, adds excessive noise, and causes pump vibration.

Diagnosing Troubles

As we said at the beginning of this chapter, many troubles in the system are wrongly blamed on the pump. A lot of pumps are removed that have nothing wrong with them, or that failed because of improper operation. So it is important to understand how the system works—where the oil goes and what happens to it—to diagnose troubles. Following are some of the problems that may be encountered and possible reasons for them.

No pressure. Remember that a pump doesn't put out pressure but flow. Pressure is caused by resistance to flow. Low pressure means the fluid is meeting little resistance. If the load does not move, the oil has probably found an easier path back to the reservoir through leakage. But remember the system must leak the full pump delivery to get any loss in pressure.

A pump usually will not lose its efficiency all at once, but gradually. So there will be a gradual slowing down of the actuator speed as the pump wears. If the loss is sudden, and the pump is not making a furious racket, best chances are the leak is somewhere else.

Slow operation. This can be caused by a worn pump or by partial leakage of the oil somewhere else in the system. There will not be a corresponding drop in pressure if the load moves at all. Therefore, horsepower is still being used and is being converted into heat at the leakage point. You can often find this point by feeling the components for unusual heat.

No delivery. If you know for certain that no oil is being pumped, it can be because the pump is not assembled correctly, is being driven in the wrong direction, has not been primed, or has a broken drive shaft. The reasons for no prime are usually improper startup, inlet restrictions, or low-oil level in the reservoir.

Noise. Any unusual noise is reason to shut a pump down immediately. Find the trouble before a lot of damage is done. Cavitation noise is caused by a restriction in the inlet line, a dirty inlet filter, or too high a drive speed.

Air in the system also causes noise. Air will severely damage a pump because there will not be enough lubrication. This can occur from low oil in the reservoir, a loose connection in the inlet, a leaking shaft seal, or no oil in the pump before starting.

Table 2.1 elaborates on the potential causes of pump failures and how you can remedy them.

Table 2.1
Diagnosing Pump Failures

Trouble	Cause	Remedy
Excessive pump noise	1. Pump-motor coupling misalignment	Realign pump & motor accurately. Align to within .005" total indicator reading.
	2. Oil level low	Fill reservoir so that surface of oil well above end of suction line during all of work cycle. About 1½ pipe diameters minimum.
	3. Pump running too fast	Reduce speed. Speeds above rating are harmful and cause early failure of pumps. Refer to pump rating for maximum speed.
	4. Wrong type of oil	Use a good, clean hydraulic oil having the viscosity in accordance with recommendations.
	5. Air leak in suction line; Air leak in case-drain line; Air leak around shaft seal.	Pour hydraulic oil or grease joints and around shaft while listening for change in sound of operation. Tighten or replace.
	6. Direction of pump rotation not correct	Arrow on pump case must agree with direction of rotation.

Trouble	Cause	Remedy
	7. Reservoirs not vented	Allow reservoir to breathe so oil level may fluctuate as required to maintain atmospheric pressure in tank.
	8. Air-bound pump	Air is locked in pumping chamber and has no way of escape. Loosen pressure line or install special bypass line back to tank so that air can pass out of the pump. An air-bleed valve need is indicated.
	9. Restricted flow through suction piping	Check suction piping and fittings to make sure full size is used throughout. Make sure suction line is not plugged with rags or other foreign material.
	10. Restricted filter or strainer	Clean filter or strainer.
	11. Worn or broken parts	Replace.
System excessively hot	1. Pump operated at higher pressures than required	Reduce pump pressure to minimum required for desired performance.
	2. Pump slippage too high	Tighten bolts on cover.
	3. Excessive friction	Internal parts may be too tight.
	4. Oil in reservoir low	Raise oil level to recommended point.
	5. System leakage excessive	Check progressively through the system for losses.

(continued)

(continued from previous page)

Trouble	Cause	Remedy
Leakage at oil seal	1. Seal installed incorrectly	Correct installation.
	2. Pressure in pump case	Observe case drain line for restriction. Check drain line circuitry for excessive back pressure arrangement.
	3. Poor coupling alignment	Realign pump & motor shafts. Align to within .005 in. total indicator reading.
	4. Packing damaged at installation; damaged or scratched shaft seal	Replace oil seal assembly. Slip packing carefully over keyway.
	5. Abrasives on pump shaft	Protect shaft from abrasive dust and foreign material.
Bearing failure	1. Abuse during coupling installation to pump	Most pumps are not designed to handle end thrusts against the drive shaft. Eliminate all end play. Couplings should be a slip fit onto the pump shaft.
	2. Excessive or shock load	Reduce operating pressure. Observe maximum rating of operating pressure. Make necessary circuit changes.
	3. Chips or other foreign matter in bearing (contamination)	Make sure clean oil is used.
	4. Coupling misalignment	Realign pump and motor.
Pump not delivering oil	1. Wrong direction of pump rotation	Observe arrow on pump case or nameplate. Direction of rotation must correspond.

Trouble	Cause	Remedy
	2. Oil level low in reservoir	Maintain oil level in reservoir well above bottom of suction line at all times.
	3. Air leak in suction line	Apply good pipe compound, nonsoluble in oil, and tighten joints.
	4. Pump running too slowly	Increase speed. Check minimum speed recommendations to be sure of proper priming.
	5. Suction filter or plugged line	Filters must be cleaned of lint or dirt soon after first start of unit. Periodic checks should be made as a preventive maintenance precaution.
	6. Bleed-off in other portion of circuit	Check for valves or other controls connected to tank.
	7. Sheared key at rotor or coupling	Check and replace.
	8. Pump cover too loose	Tighten bolts on pump cover.

Hydraulic Valves

There are three different classifications of valves:

- **Pressure control:** To regulate or limit force.
- **Directional control:** To start, stop, and reverse cylinders and motors.
- **Volume control:** To regulate rate of fluid flow or speed of an actuator.

Valves are mechanical devices consisting of a body internally bored with cylindrical chambers and passageways. The chambers may contain pistons, spools, poppets, balls, or springs. The valve may also be equipped with an adjustment screw. The arrangement of the passages in the block together with the movement of the piston or spool causes the diversion of fluid flow. As the spool moves in its chamber, some ports are closed while others are opened. Each position of the spool changes flow direction. Volume of flow and pressure are also affected by the movement of components within the valve block and by the adjustment devices.

Various methods or a combination of methods are used to activate a valve. They may be operated manually, mechanically, electrically, pneumatically, and hydraulically.

The name of the individual valve is based on its usual function. For example, a relief valve is one that relieves excess pressure.

Pressure-Control Valves

A number of pressure-control valves are used in hydraulic systems each performing a different function. These valves are *relief valves, sequence valves, counterbalance valves, pressure reducing valves,* and *unloading valves.*

Each of these valve types are described below.

Relief Valves

A relief valve is required in any hydraulic circuit that uses a positive displacement pump to protect the system against excessive pressure. If the actuator is stalled, or simply travels as far as it can go, there must be an alternate flow path for the pump's output. Otherwise pressure will instantly rise until either something breaks or the prime mover stalls.

The relief valve is connected between the pump outlet (pressure line) and tank. It is normally closed. It is set to open at a pressure somewhat higher than the load requirement and divert the pump delivery to the tank when this pressure is reached.

A relief valve can also be used to limit the torque or force output of an actuator, as in the hydraulic press or a hydrostatic transmission.

A simple relief valve (Figure 2.10) may be little more complicated in construction than a check valve. A spring-loaded ball or poppet seals against a seat to prevent flow from the inlet (pressure) port to the outlet (reservoir) port. An adjusting screw can be turned in or out to adjust the spring load, which in turn adjusts the valve's cracking pressure.

Pressure override is often a problem with a simple relief valve. When handling flow from a fair-sized pump, the valve may override several hundred psi, not only wasting power, but overloading circuit

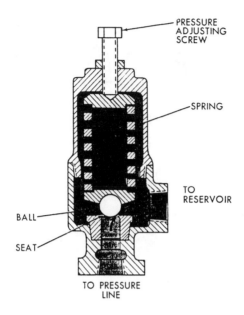

Figure 2.10. Simple relief valve.

components. Another disadvantage of this type valve is a tendency to chatter when it is "relieving." These disadvantages are overcome to a great extent in the two-stage or compound relief valve.

A compound relief valve, sometimes called a pilot-operated relief valve (Figure 2.11), is designed with a small pilot valve to limit pressure and a larger valve controlled by the pilot valve to divert the large volume of flow. Its override is low and nearly constant over a wide range of flow rates.

Some compound relief valves have a pilot stage built into a valve spool. The pilot stage is a spring-loaded poppet. The spring in the pilot stage controls the cracking pressure; the larger spring pushing against the spool determines the maximum override.

When the valve first fills through the pressure port, drilled passages carry the oil through an orifice to the spring end of the valve spool. An opening in this end of the spool leads to the head of the pilot stage.

When the passages are filled, any pressure less than the valve setting will equalize, from the pressure port to the head of the pilot-stage poppet. We then have equal pressure on both ends of the valve spool. The only effective force on the spool is the spring. It holds the spool in its normally closed position.

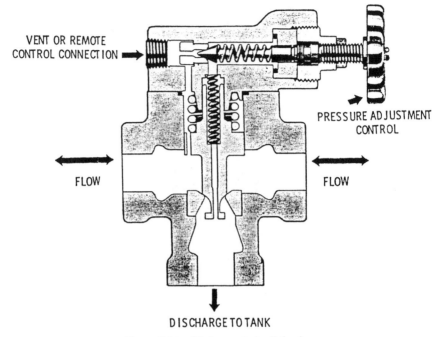

VENT OR REMOTE
CONTROL CONNECTION

PRESSURE ADJUSTMENT
CONTROL

FLOW

FLOW

DISCHARGE TO TANK

Figure 2.11. Pilot-operated relief valve.

If pressure builds up high enough to force the pilot-stage poppet off its seat, oil flows from the pressure port, through the orifice in the spool, past the pilot poppet, and through a drilled hole to the tank port.

Pilot flow causes a pressure drop across the orifice, so that pressure is no longer equal on both ends of the spool. At about a 20–40 psi pressure difference, pressure at the inlet side overcomes the spool spring. The entire spool then is pushed to open the pressure port to the tank port.

The spool assumes a position where it is balanced between system pressure and pilot-stage pressure plus the spring's force. It throttles pump delivery to the tank while maintaining pressure in the system.

When the system pressure drops, the pilot stage closes and pilot flow stops. With no flow across the orifice, pressure on the spool ends again equalizes. The spring then moves the spool back to its closed position.

Since the large spring is very light, its override is negligible. Override in the pilot stage also is slight, because it handles only a small part of the total flow, thus providing low override.

Sequence Valves

Sequence valves control the sequence of operation between two branches of a circuit. They are used to regulate the operating sequence of two separate working components in a predetermined order based on pressure signals. Fluid is directed only to that part of the system that is connected with the primary port of the valve until the pressure setting is reached. The valve then directs fluid to the secondary branch with no variation in pressure on the primary side of the system.

Sequence valves are normally closed valves. They remain closed until pressure at their inlet port reaches a pressure set by adjusting a spring. At this pressure they open and allow fluid flow through the valve, while maintaining pressure at least equal to its setting on the primary side of the system. They are similar to compound-relief valves, except that their spring chambers are drained externally to the system reservoir instead of internally to the outlet port as in relief valves.

Counterbalance Valves

Counterbalance valves are used in a circuit to maintain an adjustable resistance against flow in one direction but permit free flow in the opposite direction. The function of the valve is to prevent uncontrolled movements or to support a weight in one part of the system while fluid is made available for working components in other parts of the same system.

Counterbalance valves have built-in check valves, which allow free reverse flow through the valve. Flow is shut off at the valve until pressure at its inlet port reaches the value set on the valve's spring. Fluid then flows through the valve and returns to the system reservoir.

If counterbalance valves are set for too high an operating pressure, the system is inefficient. The correct setting is preferably 10% above the pressure setting required to compensate for a load.

Pressure-Reducing Valves

Pressure-reducing valves are used to regulate the normal operating pressure of a main circuit to the required pressure of a branch circuit. The desired reduced pressure can be obtained by adjusting the regulating device on the valve. As soon as the desired pressure is reached in the secondary system, the valve partially closes so that just enough fluid flows through to maintain the desired pressure.

Pressure-reducing valves are normally open valves. An adjustable spring holds the spool open. Reduced pressure acts on the spool to close the valve. Fluid flows unrestricted through the valve until force from the reduced pressure equals spring force. The spool moves, compressing the spring and restricting flow through the valve.

Unloading Valves

These valves are most often used in two-volume circuits, where a large quantity of oil at low pressure is required for one portion of a cycle and a relatively small amount is required at high pressure during another cycle.

Two pumps are used during low-pressure, high-speed operations, the delivery of the large one being diverted to the tank through the unloading valve when the system pressure exceeds its setting.

Remote system pressure is used to automatically open and close the valve at the proper time. Usually an accumulator, or small pump, supplies the necessary pilot pressure for actuation of the valve. An adjusting screw is provided to regulate pressure. The valve contains a compression spring that acts against the fluid-pressure force. It also contains an external pilot connection. Fluid force and spring force move the spool and direct fluid flow according to the requirements of the system.

Check Valves

A check valve (Figure 2.12) can be either a pressure control or a directional control, or both.

Often a check valve is nothing more than a ball and seat placed between two ports. As a directional control, it has a free-flow and a no-flow direction. Flow through the seat will push the ball away and permit free flow. Flow in the other direction pushes the ball against the seat, and pressure buildup forces it to seal the passage so flow is blocked.

The valve spring may be very light if it is used only to return the ball to its seat when flow stops. In that case, pressure drop through the valve will probably be no more than 5–10 psi. When the valve is used to create a back pressure, a heavier spring is used. Pressure at the inlet balances against the spring force to produce a significant pressure difference, depending on the spring rate.

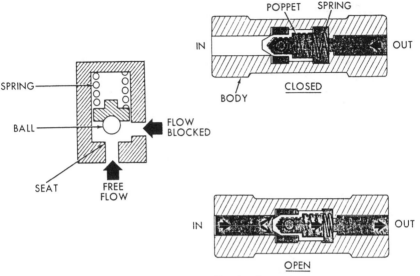

Figure 2.12. Check valves.

Directional-Control Valves

A directional-control valve (Figure 2.13) consists of a body with internal flow passages that are connected and disconnected by a movable part. This action results in the control of fluid direction. Various valve configurations are utilized depending on the component to be controlled and the control actions or positions desired. The number of passages in the valve establishes the number of *ways* that the flow can be diverted; the number of *positions* to which the valve can be shifted is determined by the configuration as well as the valve actuation mechanism (Figure 2.14).

A 2-way directional valve consists of two passages that are connected and disconnected. In one extreme spool position, the flow path through the valve is open. In the other extreme, the flow path is blocked. A 2-way directional valve gives an on-off function. This function is used in many systems to serve as a safety interlock and to isolate and connect various system parts.

A 3-way directional valve consists of three passages within a valve body—a pressure passage, one actuator passage, and one tank or exhaust passage. The function of this valve is to pressurize and drain one actuator port. When the spool of a 3-way valve is in one extreme position, the pressure passage is connected with the actuator passage.

When it is in the other extreme position, the spool connects the actuator passage with the tank line.

A 3-way directional valve is used to operate single-acting actuators such as rams and spring-return cylinders. In these applications, a 3-way valve directs pressurized fluid to one end of the cylinder. When the spool is shifted to the other extreme position, the actuator passage is connected to the tank-exhaust passage. The actuator is returned by a spring or weight.

Four-way directional valves are capable of reversing the motion of a double-acting cylinder or reversible hydraulic motor. They have a pressure passage, two actuator passages, and one tank connection. To perform the reversing function, the spool connects the pressure passage with one actuator passage. At the same time, the remaining actuator passage is connected to the tank connection by the valve spool.

The directional-valve spool can be positioned in one extreme position or the other. The spool is moved to these positions by mechanical, electrical, pneumatic, hydraulic, or manual power.

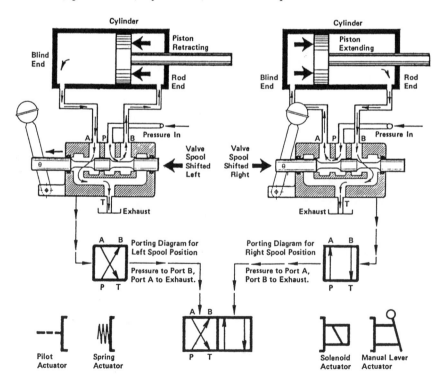

Figure 2.13. Directional-control valve operation. (Courtesy of Womack Educational Publications.)

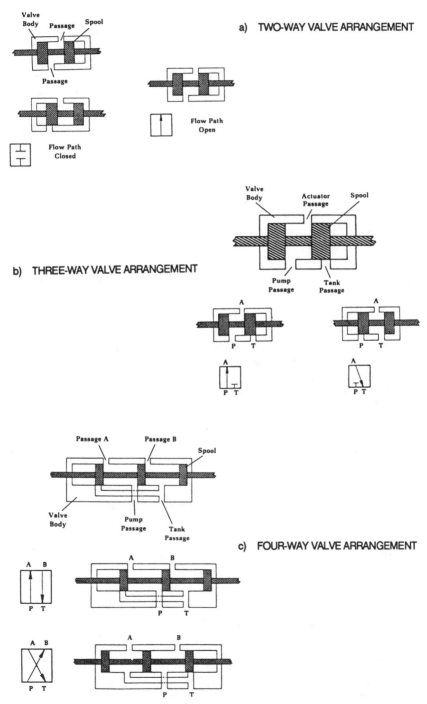

Figure 2.14. Directional-control valve arrangements.

Directional valves whose spool are moved by muscle power are known as manually operated or manually actuated valves. Various types of manual actuators include levers, push buttons, and pedals.

A very common type of mechanical actuator is a plunger. Equipped with a roller at its top, the plunger is depressed by a cam, which is attached to an actuator. Manual actuators are used on directional valves whose operation must be sequenced and controlled at an operator's discretion. Mechanical actuation is used when the shifting of a directional valve must occur at the time an actuator reaches a specific position.

Directional valve spools can also be shifted with fluid pressure, either pneumatic or hydraulic. In these valves, pilot pressure is applied to the spool ends or to separate pilot pistons.

One of the most common ways of operating a directional valve is with a solenoid. A solenoid is an electrical device that consists basically of a plunger, frame, and wire coil. The coil is wound inside the frame. The plunger is free to move inside the coil. When an electric current passes through a coil of wire, a magnetic field is generated. This magnetic field attracts the plunger and pulls it into the coil. As the plunger moves in, it contacts a pushpin and moves the directional valve spool to an extreme position.

The shifting force that can be developed by a solenoid of reasonable size is limited. In large valves, the force required to shift a spool is substantial. As a result, only smaller valves use solenoids for shifting directly. Larger hydraulic directional valves (35 gpm and larger) use pilot pressure for shifting.

In these larger directional valves, a small directional valve is often positioned to operate the larger valve. Flow and pressure from the small valve is directed to either end of the large valve spool when shifting is required. These valves are designated solenoid-controlled, pilot-operated directional valves.

Hydraulic 4-way valves are quite often three-position valves consisting of two extreme positions and a center position (Figure 2.15). The two extreme positions of 4-way directional valves are directly related to the actuator's motion. They are the power positions of the valve. The center position of the directional valve is designed to satisfy a need of the system. For this reason, a directional valve's center position is commonly referred to as a neutral condition.

There are a variety of center positions available with 4-way directional valves. Some of the more popular ones are the open center, closed

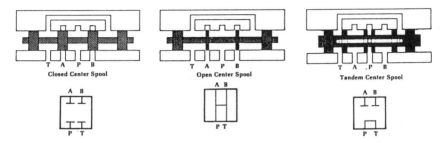

Figure 2.15. Three-position spool center configurations.

center, tandem center, and float center. These center positions can be achieved within the same valve body simply by using the appropriate spool. A directional valve with an open center spool has P, T, A, and B passages all connected to each other in the center position. An open center allows free movement of an actuator while pump flow is returned to tank. The closed center spool has P, T, A, and B passages all blocked in the center position. A closed center stops the motion of an actuator and allows each individual actuator in the system to operate independently from one power supply.

A tandem-center spool has P and T passages connected, and A and B passages blocked in the center position. A tandem-center condition stops the motion of an actuator, but allows pump flow to return to tank without going through a relief valve.

A directional valve with a float-center spool has the P passage blocked, and A, B, and T passages connected in the center position. A float-center condition allows independent operation of actuators tied to the same power source and allows free movement of each actuator.

Directional valves with three positions must have the ability to hold the spool in the center position. Spring centering is the most common means of centering a directional-valve spool. When the valve is actuated, the spool moves from the center position to one extreme, compressing a spring. When the valve is not actuated, the spring returns the spool to the center position.

Flow-Control Valves

In any hydraulic system, it is necessary to control the energy that the system delivers. This control must be capable of varying the size of the force, the direction of the force, and the speed at which the force is applied.

When flow rates must be changed in order to vary speed, the simplest method is to add a flow-control valve (Figure 2.16) in the system at an appropriate place. A flow-control valve is a device that controls flow rate through use of an orifice. In simple valves, the orifice and its associated pressure drop provide the throttling effect. In others, an orifice is used to detect changes in rate of flow, and the changing pressure drop through the orifice operates a valve member that controls the flow.

Needle Valves

One of the simplest, adjustable flow controls is the needle valve. These valves are available in all shapes and sizes, and can be purchased with or without an integral check valve.

Adding an integral check valve permits free flow in one direction and throttled or metered flow in the other. Usually, valves that have free flow capability have an arrow or some other marking on the valve body to indicate the direction of free flow, and are called flow-control valves. Without the check valve, they usually are called needle valves.

An adjustable stem changes the needle position relative to its orifice for a greater or lesser degree of throttling. Some stems have slots that accommodate screwdrivers; other valves have some type of handwheel for needle adjustment.

Pressure Compensation

Changes in the load on the actuator will cause the driving pressure to vary in accordance with the load resistance. This results in a change in the pressure difference between the flow-control valve inlet and outlet pressure. Such a change will cause the actuator speed to vary since the flow rate through the flow-control valve at a specific setting is dependent on the pressure difference (pressure drop) across the valve. Most applications do not tolerate the wide range of speed variation caused by the change in pressure difference due to actuator load variations. To provide actuator speed regulation within closer limits, pressure-compensated flow-control valves were developed. Over a somewhat limited range of pressure drop, a pressure-compensated flow-control valve uses a fixed- or manually adjusted orifice to control hydraulically the size of a second or throttling orifice. This type of valve provides a specified pressure drop as long as the inlet pressure of the valve falls within its operating limits.

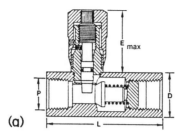

(a)

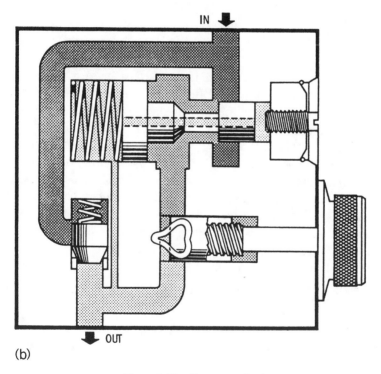

(b)

Figure 2.16. Flow-control valve.

Temperature Compensation

Changes in oil temperature alter flow rate through orifices because of changes in viscosity that occur as oil is heated or cooled. Usually, when a hydraulic system is first started, fluid temperature is lower than it will be after a period of operation. This means that not only will oil viscosity decrease as temperature increases, but system component clearances and tolerances also will change as well. These changes cause flow differences that affect the operational characteristics of the system.

To compensate for this flow-rate change, some flow-control valves use sharp-edged control orifices. These orifices are viscosity insensitive and do not alter flow rates with temperature and viscosity changes. Flow accuracies of 2% are possible with viscosity changes of as much as 250 ssu.

Another compensating method uses the phenomenon of different expansion rates of dissimilar metals. As oil temperature increases, the flow control is made smaller as a temperature compensation rod changes the size of the orifice opening. This metal expansion must match the viscosity change of the oil so that the net result is a constant flow rate.

Servicing, Repair, and Valve Troubleshooting

Hydraulic valves are precision made and must be very accurate in controlling the pressure, flow, and direction of fluid within a system. Generally, no seals or packings are used on the moving parts of the valves since leakage is slight as long as the valves are carefully fitted and kept in good condition. The clearances between the moving parts are very small—on the order of 0.0002 in. for some typical spool-type valves.

Contaminants, such as dirt in the oil, are the major causes for valve failures. Small amounts of dirt, lint, rust, or sludge can cause annoying malfunctions and extensively damage critical valve parts. Such contaminants will cause the valve to stick, plug small openings, or scratch the mating surfaces until the valve leaks. Any of these conditions will result in poor system operation, or even complete stoppage. This damage can be eliminated if the operators use some care in keeping dirt out of the system. Some of the rules to follow for reducing valve problems are:

- Use the oil that is specified for the system—make no substitutions.
- Follow the recommendations in the operator's manual.
- Use an oil that will not oxidize or prevent generation of rust particles.
- Change the oil and the filters on a regular basis.

In order to perform service work on the valves used in a system, it is often necessary to remove them. The following precautions should be taken before disconnecting or removing valves:

1. Disconnect the electrical power source before removing the valve. This will eliminate accidental starting or movement; tools can

short out the circuit when coming into contact with a "hot" electrical system.

2. Move the valve-control lever in all directions in order to release any hydraulic pressure in the system before disconnecting any valves.

3. Either block up or lower all hydraulic working units before disconnecting any valves or fittings. This will assure that the actuator will not drop, possibly causing harm to the maintenance person.

4. Clean the valve and the surrounding area before removing any fittings or the valve. Do not allow water or other contaminants to enter the system. Make sure that all hose and line fittings are tight before starting the machine.

5. Use fuel oil or other suitable cleaning solvent to remove grease and dirt; never use a paint thinner or acetone as a cleaning agent. Plug all portholes immediately after disconnecting the lines in order to prevent dirt and moisture from entering the system.

Valve Disassembly Hints

In order to determine the problem with a valve assembly, it is often necessary to disassemble the unit. During disassembly, some precautions are necessary in order to prevent damage to the valve parts. It should also be kept in mind that, during the disassembly operation, all parts are to be inspected for the purpose of obtaining clues pointing to the cause of the failure experienced. Following are some hints that will lead to the successful disassembly and repair of the valve assemblies:

1. Do not perform hydraulic-valve internal service work in the shop, on the ground, or where there is danger of dirt or dust being blown into the valve parts. Use only clean work areas. Be sure that all tools are clean and free of grease, dirt, and grime.

2. During disassembly of the valves, be careful to identify the parts for reassembly. Spools are selectively fitted to the valve bodies and must be reinstalled in the same bodies from which they were removed—and in the same position. Valve sections in stack-type assemblies must be reassembled in the same order in which they were.

3. When it is necessary to clamp a valve housing in a vise, be sure to use extreme caution in terms of how hard the part is clamped, to prevent damage to the component. If possible, use a vise with

lead or brass jaws or protect the part by wrapping it in a protective covering of some sort.

4. All valve-housing openings should be sealed when components are removed during disassembly. This will prevent foreign material from entering the valve housing.

5. Many valve assemblies have internal springs to center or spring-load the valve. On such valves, be very careful when removing the cover plate or backup plug since personal injury may result from the spring tension being suddenly released. When springs are under high preload, use a press to remove them. This will enable the tension to be released in a gradual and controlled manner.

6. As the valve assembly is taken apart, wash all of the valve components that are removed in a clean mineral oil solvent (or any other noncorrosive-type cleaner). Dry the parts with compressed air and place them on a clean surface for inspection. Do not wipe the parts with rags or wastepaper. Lint deposited on any parts can enter the hydraulic system and cause trouble.

7. After the parts are cleaned and dried, coat them immediately with a rust-inhibiting hydraulic oil. Then be sure that the parts are kept in a clean place free from moisture until they are reinstalled.

8. Be sure to carefully inspect all valve springs during disassembly. All springs that show signs of being crooked or cocked, or contain broken, rusty, or fractured coils are to be replaced.

Valve Repair

Directional-control valve repair. Directional valve spools are installed in the valve housing by a select fit process. This is done in order to provide for the closest possible fit between the housing and the spool. The closer this fit is, the less the internal leakage, and the valve will also have maximum load-holding qualities. In order to achieve such a close fit, special factory techniques and equipment are required. Therefore, most valve spools and housings are furnished for service only in matched sets. They are not available from the manufacturer as individual parts for replacement.

When repairing a directional valve assembly, inspect the valve spools and bores for burrs and scoring. Spools may become coated with impurities from the hydraulic oil. Figure 2.17 depicts the inspections to be performed.

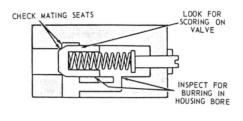

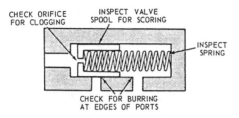

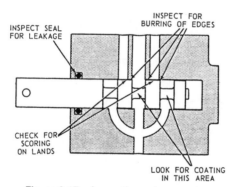

Figure 2.17. Inspecting valve spools.

In cases where the scoring and coating is not deep enough to cause objectionable leakage, the surfaces can be polished with crocus cloth. Be sure not to remove any of the valve material—only the coating is to be removed. Replace the valve body and spool if the scoring or coating are excessive. If the action of the valve was erratic or sticky before removal from the system, it may be unbalanced due to wear on the spool or body. In this case, the assembly should be replaced.

O-rings are used on many directional valves as seals. These should be inspected for possible damage. Figure 2.18 depicts some typical O-ring failures and their causes. Prior to reinstalling a new O-ring, the cause of the failure must be corrected. This may involve selection of a different material, being more careful in the installation process, or correcting hydraulic system problems.

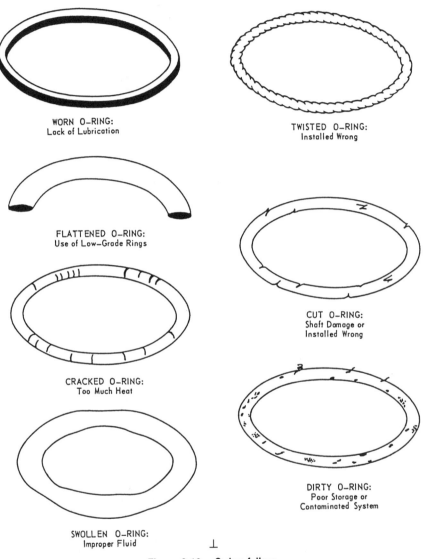

WORN O-RING:
Lack of Lubrication

TWISTED O-RING:
Installed Wrong

FLATTENED O-RING:
Use of Low-Grade Rings

CUT O-RING:
Shaft Damage or
Installed Wrong

CRACKED O-RING:
Too Much Heat

DIRTY O-RING:
Poor Storage or
Contaminated System

SWOLLEN O-RING:
Improper Fluid

Figure 2.18. O-ring failure.

The valve spool should be checked for freedom of movement in the housing bore. When lightly oiled, the spool should slide into the bore from its own weight. If some force is necessary to move the spool, then the fit is not proper and the valve assembly should be reinspected.

Flow-control valve repair. Some flow-control valves have spools with internal orifices. These orifices may become clogged with dirt or other foreign matter. Thus, the valve spools should be inspected for clogging and cleaned with compressed air or a small-diameter wire. Again, caution must be taken so that contaminants are not introduced during the inspection and cleaning process. Figure 2.17 depicts the inspections to be performed on nonadjustable, fixed-type flow-control valve assemblies. The movable members in the valve should be inspected for scoring, wear, or coating. Also, the springs in the valve assembly should be inspected. Any problems noted should be corrected in a similar fashion to that described for directional valves.

Needle-type flow-control valves should be inspected for damage to the tapered section on the valve stem; also, the valve seat should be inspected for erosion damage. Slight damage can be corrected by polishing the marks from the valve stem with crocus cloth or very fine emery paper. Rewash all parts thoroughly to remove all emery or metal particles. Any such abrasive materials could quickly damage the entire hydraulic system.

Pressure-control valve repair. Pressure-control valves are force-balance–type devices in that a spring force is counteracted by a hydraulic force. Thus, it is very important that the springs are proper. Check for weak springs with a spring tester if system checks have indicated low-pressure readings. This can be remedied by replacing the spring or by adding shims to increase the compression of the spring in some cases. Never add so many shims that the spring is compressed solid.

The second area of concern with pressure-control valves is the condition of the valve seats and poppets. These are the components that prevent leakage when the valve is closed. Again, refer to Figure 2.17 for the inspections that are to be made.

Check the pressure-control valve seats for evidence of scoring or leakage. The valve should be replaced if flat spots appear on the poppet or on the seat. If the scoring is not very deep, then metal valve seats and poppets can be surface polished using crocus cloth. Care should be taken not to remove any valve material—that is, the size of the original component should not be changed.

Some seats and poppets are made of plastic materials such as nylon. These materials are long wearing and they are sufficiently elastic to conform perfectly to mating surfaces, resulting in a tight seal. The plastic seats and poppets will wear but the mating metal surface will have no damage. When repairing this type of valve, always replace the plastic parts with new ones.

Valve Assembly Hints

Just as was the case during disassembly of valves, some precautions are necessary during the assembly process. Following is a listing of some considerations to be taken during the valve assembly process:

1. When assembling valves, be sure that they are kept absolutely clean. Wash all of the parts in a suitable solvent, blow dry with compressed air, then dipped in hydraulic oil that contains a rust inhibitor to prevent rusting. This process will also help in assembly and will provide some lubrication. Petroleum jelly can also be used to hold seal rings in place during the assembly process.
2. Check all of the valve parts again to be sure that the valve mating surfaces are free of burrs, scratches, paint, etc.
3. Replace all seals and gaskets with new ones when repairing the valve assembly. Soak the new seals and gaskets in clean hydraulic oil before assembly. This will prevent damage and help seal the valve parts.
4. The valve spools must be inserted into their matched bores. Care must be taken to assure that this step is followed. Also, valve sections on stack-type valve assemblies must be reassembled in their correct order. Failure to do so will result in faulty system operation.
5. When mounting valves on the machine, care must be taken so that there is no distortion. This may be caused by uneven tension on the mounting bolts, improper tightening of the bolts, uneven mounting surfaces, improper location of the valve assembly, or insufficient allowance for expansion when the oil temperature rises during system operation. Any of these can result in valve-spool binding.
6. Following assembly and installation (mounting) of the valve, check the action of the valve spools. If there is any sticking or binding, then the tension on the mounting bolts must be checked along with all of the other possible causes of binding listed previously.

Hydraulic Valve Troubleshooting

Troubleshooting problems with hydraulic valves requires that some basic steps be followed. These steps will help to diagnose most hydraulic valve difficulties. Listed in Table 2.2 are some troubleshooting steps that may be used in the diagnosis of valve problems.

(text continued on page 83)

Table 2.2
Diagnosing Valve Problems

Pressure-Control Valves Relief Valves	
Trouble	Cause
Low or erratic pressure	1. Incorrect valve adjustment 2. Dirt, lint, or burr holding valve partially open 3. Worn or damaged poppet or seat 4. Sticking valve piston in main body 5. Weak spring 6. Spring ends damaged 7. Valve piston or poppet cocking in body or on seat 8. Orifice or balance hole blocked
No pressure	1. Orifice or balance hole plugged 2. Poppet not seating or seat damaged 3. Loose fit between body and piston 4. Valve piston binding in body or cover 5. Spring broken 6. Dirt or burr holding the valve partially open 7. Worn or damaged poppet or seat 8. Valve piston or poppet cocked in body or seat
Excessive noise or chatter	1. Oil viscosity too high 2. Damaged or worn poppet or seat 3. Excessive return-line pressure 4. Pressure setting too close to that of another valve in the circuit 5. Improper spring used in the valve
Unable to adjust properly without getting excessive system pressure	1. Spring broken 2. Spring fatigued 3. Incorrect spring 4. Drain-line or return-line restricted

(continued)

(continued from previous page)

Pressure-Control Valves Relief Valves	
Trouble	**Cause**
Overheating of system	1. Continuous operation at relief-valve setting 2. Oil viscosity too high 3. Internal leakage in valve at seat

Pressure-Reducing Valves	
Trouble	**Cause**
Erratic pressure	1. Dirt in the oil 2. Worn poppet or seat 3. Restricted orifice or balance hole 4. Valve spool binding in the body 5. Drain line not open freely to the reservoir 6. Spring ends not square 7. Incorrect spring 8. Fatigued spring 9. Valve requires adjustment 10. Worn spool or housing bore
No pressure control	1. Drain line plugged or restricted 2. Valve spool stuck in bore 3. Broken spring

Sequence Valves	
Trouble	**Cause**
Premature movement to secondary operation	1. Valve setting too low 2. Excessive load on the primary cylinder 3. High inertia load on the primary cylinder 4. Primary cylinder damaged
No movement or slowness of secondary operation	1. Valve setting too high 2. Relief-valve setting too close to that of the sequence valve 3. Valve spool binding in the body

Sequence Valves	
Trouble	Cause
Valve not functioning properly	1. Improper installation 2. Incorrect adjustment 3. Broken spring 4. Dirt or foreign material on plunger seat or in orifices 5. Leaky or blown gasket 6. Drain line plugged 7. Valve covers not tightened properly or installed incorrectly 8. Valve plunger worn or scored 9. Seat of valve stem scored or worn 10. Orifices too large, causing jerky operation 11. Binding due to coating on moving parts of valve from oil impurities (due to overheating of oil or incorrect oil)

Unloading Valves	
Trouble	Cause
Valve fails to completely unload the pump	1. Valve setting too high 2. Pump failing to build up to unloading-valve pressure 3. Valve spool binding in the body

Direction-Control Valves	
Trouble	Cause
Faulty or incomplete shifting	1. Worn or binding control linkage 2. Insufficient pilot pressure 3. Burned out or bad solenoid 4. Defective centering spring 5. Improper valve-spool adjustment
Cylinder Creeps or Drifts	1. Valve spool is not centering properly 2. Valve spool has not shifted completely

(continued)

(continued from previous page)

Direction-Control Valves	
Trouble	Cause
	3. Valve spool or body are worn 4. Valve seats leaking 5. Leakage past piston seals in the cylinder
Cylinder load drops with valve spool in centered position	1. Loose lines between cylinder and valve 2. O-rings on lockout or holding valve (if used) leaking or damaged 3. Broken springs in holding or lockout valves 4. Relief valve leaking
Cylinder drops load slightly when raised	1. Spool valve position improperly checked 2. Improper adjustment of valve-spool position

Flow-Control Valves	
Trouble	Cause
Variation in flow	1. Valve spool binding in body 2. Oil viscosity too high 3. Leakage in cylinder or hydraulic motor 4. Dirt in the oil 5. Insufficient pressure drop across the valve
Improper flow	1. Valve not adjusted properly 2. Restricted valve-spool travel 3. Restricted passages or orifices 4. Cocked valve spool 5. Circuit-relief valve leaking 6. Oil temperature too high
Erratic pressure	1. Worn valve plunger or seat 2. Dirty oil

(text continued from page 78)

Cylinders

In all applications, fluid energy must be converted to mechanical energy before any useful work can be done. Cylinders convert the fluid-power energy into straight-line mechanical energy.

A cylinder consists of a cylinder body, a piston, and piston rod attached to the piston. End caps are attached to the cylinder body with threaded tie rods, or they are welded. As the cylinder rod extends and retracts, it is guided and supported by a bushing called a rod gland. The side through which the rod sticks out is called the "head." The opposite side without the rod is termed the "cap." Inlet and outlet ports are located in the head and cap ends. Cylinder construction is shown on Figure 2.19.

For efficient operation, a leak-free seal must exist across the cylinder's piston as well as at the rod gland. Hydraulic cylinders often use cast-iron piston rings as a piston seal. Piston rings are a durable seal but they have some leakage flow due to the clearances between the piston ring and cylinder-body tube.

Hydraulic systems, which cannot tolerate any leakage flow, use a resilient piston seal. Resilient seals do not leak under normal conditions, but are less durable than piston rings.

Rod-gland seals come in several varieties and are generally resilient seals. Some cylinders are equipped with U, V, multilip, or cupped-shape primary seal and a rod wiper, which prevents foreign materials from being drawn into the cylinder. Material used in the cylinder parts are detailed in Table 2.3; various types of packing and cylinder seals are listed in Table 2.4.

(text continued on page 86)

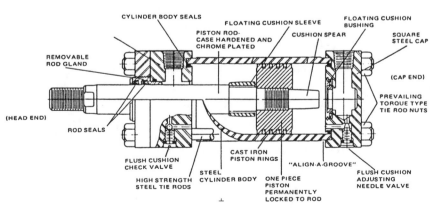

Figure 2.19. Cylinder construction.

Table 2.3
Materials Used in Cylinder Parts

Cylinder Part	Service Medium		
	Hydraulic	Air	Water

Covers	Steel bar
	High-tensile
	  cast iron
	Cast steel
	Forged steel
		Steel bar
		High-tensile
		  cast iron
		Aluminum
		Brass
			Bronze
			Plated steel

Tube	Steel
	Chrome-plated
	  steel
		Brass
		Chrome-plated
		  steel
		Aluminum-
		  coated
		Reinforced
		  plastic
			Bronze
			Chrome-plated
			  steel

Piston rod	Steel
	Chrome-plated
	  steel
	Hardened and
	  chrome-plated
	  steel
		Steel
		Chrome-plated
		  steel
			Stainless steel
			Chrome-plated
			  stainless steel

Pistons	Cast iron Bronze-faced steel	Cast iron Synthetic	Bronze

Packings
  and
  seals
	Cast iron
	  (automotive-
	  type piston
	  rings)
	"O" rings
	Quad rings
	Chevron rings
	
	Vee-type packing
	Hat-type packing
	Cup-type
	  packing
		Block vee
		
		Cup-type packing
		"O" rings
		Quad rings
		
		Hat-type packing
		"V"-cup
			Bronze
			  (automotive-
			  type piston
			  rings)
			
			Cup-type
			  packings
			Vee-type rings

Table 2.4
Various Types of Packing Materials

Material	Compatibility	Temp. Range
Leather	Good for water or oil	−65° to 200°F
Impregnated poromeric material (Corfam)	Good for water, oils, fuels, greases; not recommended for phosphate esters.	−65° to 250°F
EP rubber (ethylene-propylene)	Good for water, air, steam, phosphate esters.	−65° to 300°F
Nitrile rubber (NBR, Buna-N)	Good for water, aliphatic-base petroleum oils, and some synthetics.	−65° to 250°F
Butyl rubber	Good for water and some synthetics; not good with petroleum-based oils.	−65° to 250°F
Polyacrylates (Hycar, Vyram)	Good with some synthetics; not good with water-base fluids.	0° to 350°F
Silicone rubbers	Good for water; fair with petroleum oils; good with some synthetics.	−120° to 500°F
Fluorosilicon	Good with water, oils, and some synthetics.	−120° to 500°F
Urethane	Good for air, oil, or water.	−40° to 212°F
Fluoro-elastomers (Viton-A)	Good for water, petroleum-based fluids, and most synthetics.	−20° to 500°F
Fluoroplastics (Teflon, Kel-F)	Good for all fluids.	−320° to 500°F

(text continued from page 83)

As a cylinder piston completes its stroke, the piston runs into a cylinder end. If inertia is high enough at this point, the cylinder may experience a shock or concussion, which could be damaging. As protection, cylinders can be equipped with cushions. Cushions slow down a cylinder's piston movement just before reaching the end of the stroke. Cushions can be applied at either or both ends of a cylinder.

A cushion consists of a needle-valve flow control and a plug attached to the piston. The plug can be on the rod side, in which case it is called a cushion sleeve. Or, it can be on the cap-end side, in which case it is called a cushion spear. As a cylinder piston approaches the end of its travel, the plug blocks the normal exit for a fluid and forces it to pass through a needle-valve flow control. The needle valve restricts flow out of the cylinder slowing down piston movement. The opening of the needle valve determines the rate of deceleration. In the reverse direction, flow bypasses the needle valve by means of a check valve within the cylinder.

Along with being a bearing, a rod-gland bushing is also a fulcrum for the piston rod. If the load attached to the piston rod of a long stroke cylinder is not rigidly guided, then at full extension, the rod will tend to pivot or jackknife at the bushing, causing excessive loading. A stop tube is used to protect the rod-gland bushing, by distributing any loading at full extension between the piston and bushing. A stop tube is a solid, metal collar that fits over the piston rod. A stop tube keeps the piston and rod-gland bushing separated when a long stroke cylinder is fully extended.

Cylinders can be mounted in a variety of ways, among which are flange, trunnion, side lug and side tapped, clevis, tie rod, and bolt mounting (see Figure 2.20).

A number of common types of cylinders are used. Following is a listing of these along with brief descriptions:

- *Single-Acting Cylinder:* A cylinder in which fluid pressure is applied to the movable element in only one direction.
- *Spring-Return Cylinder:* A cylinder in which a spring returns the piston assembly.
- *Ram Cylinder:* A cylinder in which the movable element has the same cross-sectional area as the piston rod.
- *Double-Acting Cylinder:* A cylinder in which fluid pressure is applied to the movable element in two directions.
- *Single-Rod Cylinder:* A cylinder with a piston rod extending from one end.

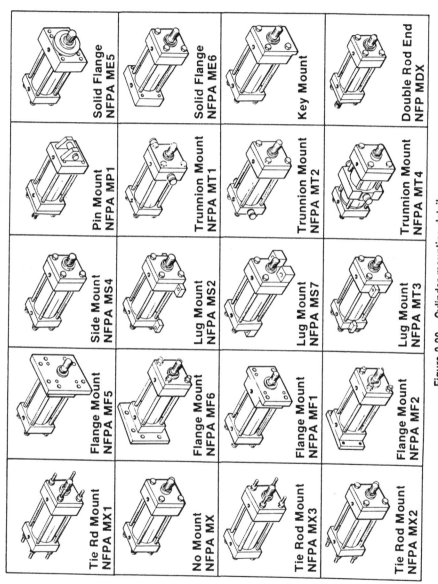

Figure 2.20. Cylinder mounting details.

Tie Rd Mount NFPA MX1

Flange Mount NFPA MF5

Side Mount NFPA MS4

Pin Mount NFPA MP1

Solid Flange NFPA ME5

No Mount NFPA MX

Flange Mount NFPA MF6

Lug Mount NFPA MS2

Trunnion Mount NFPA MT1

Solid Flange NFPA ME6

Tie Rod Mount NFPA MX3

Flange Mount NFPA MF1

Lug Mount NFPA MS7

Trunnion Mount NFPA MT2

Key Mount

Tie Rod Mount NFPA MX2

Flange Mount NFPA MF2

Lug Mount NFPA MT3

Trunnion Mount NFPA MT4

Double Rod End NFP MDX

87

- *Double-Rod Cylinder:* A cylinder with a single piston and a piston rod extending from each end.
- *Telescoping Cylinder:* A cylinder with multiple tubular-rod segments, which provide a long working stroke in a short retracted envelope. The tubular-rod segments fit into each other to provide the telescoping action.
- *Tandem Cylinder:* Consists of two or more cylinders mounted in line with pistons connected by a common piston rod. These cylinders provide increased output force when the bore size of a cylinder is limited, but not its stroke.
- *Duplex Cylinder:* Consists of two cylinders mounted in line with pistons not connected. Duplex cylinders give a 3-position capability.

Cylinder Ratings

The ratings of a cylinder include its size and pressure capability. Most come with a standard rod size although intermediate and heavy-duty rods are available. Cylinder size is piston diameter and stroke length. The speed of the cylinder, the output force available and the pressure required for a given load all depend on the piston area (.7854 multiplied by the diameter squared). The area of the piston rod must be subtracted when the piston is being retracted.

To find the speed of a cylinder when size and gpm delivery are known:

$$\text{Speed (in./min)} = \text{Flow rate (gpm)} \times \frac{231}{\text{effective piston area (in.}^2)}$$

To find the flow required for a given speed:

$$\text{Flow rate (gpm)} = \frac{\text{effective piston area (in.}^2) \times \text{speed (in./min)}}{231}$$

To find the force output for a given pressure:

$$\text{force (lb)} = \text{pressure (psi)} \times \text{effective piston area (in.}^2)$$

To find the pressure required to exert a given force:

$$\text{pressure (psi)} = \frac{\text{force (lb)}}{\text{effective piston area (in.}^2)}$$

Installation and Maintenance of Cylinders

Cylinders of even the highest quality can fail after a short period of operation if they have been poorly installed or improperly maintained. Dirt and heat are their biggest enemies. Here is a real-life example: A group of expensive heavy-duty mill cylinders were installed in a large system; after a short time they became sluggish, then failed completely. When they were dismantled, it was found that they contained sand and welding beads. An investigation disclosed that the group of cylinders to which all the failed cylinders belonged had been shipped to the erection site before the building was completed. Workmen had then removed the pipe plugs from the cylinder ports to use elsewhere on the project. That left the interior of the cylinders exposed to sand from the surrounding area and welding beads from the erection of the building frame.

When it is necessary to store cylinders at an erection site prior to installation, five simple steps will prevent maintenance headaches later:

1. Check all the pipe ports to make certain they have protective enclosures or plugs.
2. Make certain the exposed parts of piston rods, shafts, and bearings are given a protective coating or a protective cover if the units will be stored out of doors.
3. Fill hydraulic cylinders and rotary actuators with a fluid that will be compatible with that used in the system in which they will be installed. Air cylinders are often stored successfully with desiccant packs attached to the pipe ports. They may also be filled with a good grade of mineral oil; it will not harm the seals, yet will provide ample protection. The fluid should be drained completely before the cylinders are installed.
4. Keep all force components away from heat, which will age their seals and thus reduce their effectiveness.
5. Store cylinders vertically whenever possible, with the piston rod up and in the retracted position. This practice relieves the seals of the weight of the rod and the piston, and also assures protection to that part of the rod that will slide through the rod seal when the cylinder is in use.

Installation

In planning mountings, the plant engineer should bear in mind the truism that force components exert high forces. For example, a

6-in.-bore pneumatic cylinder operated at 90 psi generates a thrust of 2545 lb, and a 6-in. hydraulic cylinder operated at 3000 psi generates 84,823 lb. A hydraulic motor with a 3000-psi input can produce several hundred horsepower with torques of 20,000 ft-lb or more. When such forces are being applied, it is important not only to select and install the proper component for the job, but to install it on a structure that will stand up under all operating conditions.

Eccentric loading (side loading) should be minimized in cylinder installations. It is not only harmful to the piston rod seal and the rod bearing, but it can also cause problems with the cushioning, the piston rod, the piston, and the cylinder tube. Bent rods and scored rods are often the result of eccentric loading.

Figure 2.21 shows one method of combating eccentricity. Long bearings have been installed on the machine table to carry the eccentric load, leaving the cylinder free to perform as its manufacturer intended without undue stress. A swivel connection between the cylinder and the table would provide further protection against the eccentric load.

Figure 2.22 illustrates another method. In this setup, long rods are run through pillow blocks connected to the cylinder covers. The alignment of the blocks between the front and rear covers must be precise. The rods are connected to the pusher plate, and are guided by the bearings in the pillow blocks. Smaller eccentric loads might be countered by the use of longer-than-standard rod bearings, large-diameter piston rods, stop tubes, or cylinders with double-ended rods.

Whenever eccentric loads are present, the piston rod should have some kind of support, but it is especially desirable for long-stroke cylinders. For example, the 5¾-in.-dia rod of an 8-in. cylinder with a 12-ft stroke weighs 1,150 lb. Imagine the load applied to one side of the rod bearing and to the rod seal if the rod is extended horizontally without external support. Even if such a rod is supported, it is not advisable to allow the piston to contact the rod-end cover. A stop tube will provide additional support at the piston end of the rod, Figure 2.23.

Piston rod-seal protection is important wherever process heat may be encountered. High-temperature seals are commercially available that will withstand temperatures as great as 500°F. Another possibility is the use of heat shields or heat barriers, which will protect the fluid in the cylinders as well as the seals. In general, fluid temperatures should not be permitted to exceed 150°F; heat exchangers should be installed if system temperatures become excessive.

A rod wiper, Figure 2.24, will prevent moderate amounts of dust, grit, or liquid spatter from entering the cylinder. If the contamination

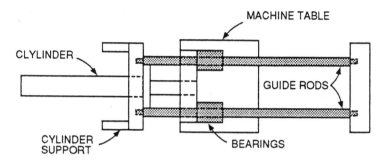

Figure 2.21. Long, sliding bearings and external guide rods provide a certain, though elaborate, cure for severe cases of eccentric loading.

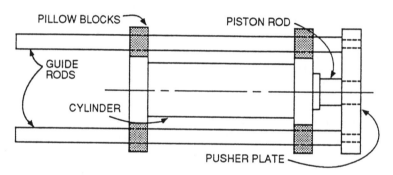

Figure 2.22.   Pillow blocks bolted to the cylinder's covers permit the guide rods to become a more integral part of the cylinder assembly.

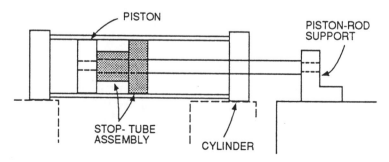

Figure 2.23.   Long cylinders with heavy rods will benefit from the use of a stop tube even when the threaded end of the piston rod is well supported.

Figure 2.24. Rod wiper and rod seal.

is too adhesive to be removed by a wiper, a scraper may be necessary; it will scrape off the contaminant as the rod is retracted, but will not score a rod with a good, hard finish. The piston rod should be hard-chrome plated.

Foot, side-lug, or centerline-lug-mounted cylinders should be solidly fastened to firm, flat supports. Because the thrust is carried at the center-line of the cylinder, the centerline-lug mounting is one of the best fixed mounts, but it is less popular than the other two because it is not as easy to use. Figure 2.25 shows a side-lug mount in which the mounting blocks are placed alongside the mounting lugs of the cylinder. The blocks should be positioned to take the force of the cylinder's load. Thus, if the load is on the forward stroke, they should be behind the lugs of the rear cover; if the load is in tension, they should be in front of the lugs of the front cover. Mounting blocks should never be put at each end of the cylinder; the shock-absorbing capabilities of the cylinder will be lost, and changes in temperature and pressure will put stress on the tube. For applications in which it is anticipated that maxi-mum force will be required for both the forward and return strokes of the cylinder, the mounting blocks should be positioned before and behind one set of lugs—usually the front cover lugs. A keyed section of the packing-retainer plate can also be used to absorb thrust.

Dowel pins can be fitted into the mounting lugs to absorb shock and assist in aligning the cylinder. They should never be put in opposite

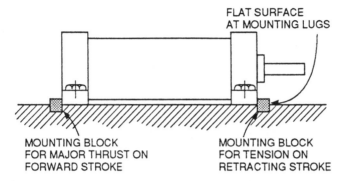

FLAT SURFACE
AT MOUNTING LUGS

MOUNTING BLOCK
FOR MAJOR THRUST ON
FORWARD STROKE

MOUNTING BLOCK
FOR TENSION ON
RETRACTING STROKE

Figure 2.25. Support for foot-mount cylinder.

lugs since this positioning will cause misalignment and internal stress when the cylinder operates at high pressures or with shock loads.

A side-lug, centerline-lug, or foot-mounted cylinder should have a center support if the stroke is long. Sagging often causes vibration and other problems.

Front-flange or rear-flange-mounted cylinders with long strokes should be installed with provision for support of the end opposite the mounting. Such support will keep the cylinder tube in alignment so that cushioning will function properly and will reduce wear on the piston, rod bearing, and seals.

When a front-end, flange-mounted cylinder is to be used as the main cylinder in a hydraulic press, only the type in which the flange is the full thickness of the cylinder cover should be used, and it should be reverse mounted. This mounting takes the stress off the bolts, and makes it easy to service the rod seal.

Trunnion-mounted cylinder applications must be furnished with solidly built trunnion supports or pillow blocks. The holes in these supports that will accept the trunnions should be 0.002-in. or 0.003-in. oversize. It is mandatory that they be in line.

Clevis-mounted cylinders find wide use wherever a pivoting action in one plane is needed. Significant side loading must be eliminated, and proper alignment to the direction of desired motion is important. When trunnion-mounted or clevis-mounted cylinders are installed, hose should be used to make the connections, so that the cylinders will be free to move without interference.

Ports tapped for American (National) Standard Dryseal Pipe Threads require the use of a sealing compound. Teflon tape is excellent. Regardless of whether tape or a paste sealer is used, it should not be applied to the first couple of threads on the pipe or fitting; otherwise, it will enter the cylinder and contaminate the system.

Hydraulic cylinders with piston rods having a cross-sectional area half as large as their piston's should have ports in the rear cover of sufficient size to allow the fluid to flow in and out without undue restriction. In general, a fluid velocity of 15 ft/sec should not be exceeded. It may be necessary to use more than one port; an oversize port is not the answer unless it is unrestricted all the way into the cylinder. Keep in mind that, because the rod occupies so much of the volume at the front of these cylinders, the return stroke can be twice as fast as the extension stroke, provided the rear connections will permit the fluid there to move out rapidly enough.

Maintenance

The preventive procedures applicable to fluid-power-force components consist largely of providing them with clean fluid and shielding them from heat and airborne dirt. Fluid motors are least tolerant of dirty fluid, but all force components will benefit from good filtration. Self-lubricating air cylinders are available, but most air cylinders require an oil mist in their air. The lubricator should be as close to the cylinder as possible, and downstream of any filter.

Cylinder rod seals can often be replaced without removing the cylinder from the machine. Many cylinders have rod-seal cartridges that are threaded into the front cover so the tie rods need not be disturbed. The maintenance mechanic installing the new seal should make certain the rod is free of anything that could cut the new seal, and spread a little light grease on the ID of the seal. The piston rod should be pulled out far enough that the wrench flats on the end of the rod will be beyond the end of the seal cartridge when it starts to enter the cylinder cover. This arrangement will allow the seal to rotate on the full circumference of the rod.

Repairs requiring disassembly should be carried out in a clean, well-lighted area. This point cannot be stressed too much, as so often maintenance is done on a cluttered workbench where small parts are easily lost.

All of the parts removed from the component undergoing repair should be thoroughly cleaned, and kept protected from dirt. If they are not to be immediately reassembled into the component, they should be coated with a protective substance or put into an oil bath.

Scored cylinder tubes can sometimes be polished with a fine emery cloth; if the scored section is deep, it may be less expensive to buy a new tube than to try to hone the old one and build it up with chrome plate.

Automotive-type piston seals (rings) should be examined carefully for cracks and score marks.

Piston rods should be checked for imperfections that might damage the rod seal—flaking chrome plating is a common problem. Rods should also be checked for straightness, since eccentric loading sometimes causes them to bend slightly.

Cushion noses and cushion recesses (in the covers) should be examined for score marks. The ball check and the needle for the cushion mechanism should also be checked for dirt deposits and for score marks. The ball-check seat can usually be polished out to form a new seat, and the nose of the needle can be refinished, if necessary.

Installing a complete repair kit instead of merely replacing a worn seal can save labor and downtime later. A repair kit usually consists of replacements for all the soft seals, a new rod bearing, and springs for the ball-check assembly, as well as any other small part the manufacturer deems necessary. It is much less expensive to install the complete kit than to replace one seal now and another later, but it is surprising how many users will buy one seal to repair a two-seal piston in a cylinder.

When a cylinder is reassembled, the mechanic should make certain that all the cover screws on both covers are torqued to the manufacturer's recommended level, and that the mounting surfaces on side-lug, foot, and centerline-mounted cylinders are in line. These should be checked on a flat surface. Before a cylinder is put back into operation, it should be bench-tested to make certain that the piston rod moves freely under pressure and that it does not leak or make unusual noises that might indicate that the parts are binding.

Start-up

Before connecting the cylinder, the hydraulic system should be thoroughly flushed. The cylinder connection lines should be shut off during this process. Thereafter, the cylinders should be first connected to the pipe system.

Before start-up, the cylinder should be vented to remove trapped air at both ends if necessary. This may be done by loosening the connections or, alternatively, by means of special bleed screws. After thorough venting (the oil must be free of bubbles and there must be no further foaming) the cylinder connections are retightened. Particularly after assembly and start-up of a new system, the cylinders should be checked at frequent intervals for correct function and for leakage.

Troubleshooting Cylinders

Problems with cylinders fall into two general categories—erratic action and the cylinder failing to move the load. There are several other problems such as leakage, seal wear, and drifting. These problems are all interrelated. Table 2.5 details each problem area, the probable causes, and remedies.

Hydraulic Motors

A hydraulic motor works in reverse when compared to a pump (Figure 2.26). The pump drives its fluid, while the motor is driven by its fluid. Thus:

- **Pump**—draws in fluid and pushes it out, converting mechanical force into fluid force.
- **Motor**—fluid is forced in and exhausted out, converting fluid force into mechanical force.

In use, the pump and motor are often hydraulically coupled to provide a power drive:

1. The pump is driven mechanically, drawing in fluid and pumping it to the motor.
2. The motor is driven by the fluid from the pump and so drives its load by a mechanical link.

The motor is really an actuator, like the cylinder. However, the motor is a rotary actuator that rotates in a full circle. The motor is designed much like the pump. Both use the same basic types—gear, vane, and piston. Often their parts can be substituted one for the other.

Both pump and motor use an internal sealing of parts to back up their flow of fluid—positive displacement. Without this seal, a motor's elements would not move under force of the incoming fluid (Figure 2.27).

(text continued on page 100)

Table 2.5
Diagnosing Cylinder Problems

Trouble	Cause	Remedy
Cylinder drifts	1. Piston seal leaks	Pressurize one side of the cylinder piston, and disconnect fluid at opposite port. Observe leakage. One to three in.^{3}/min is considered normal for piston ring leakage. Virtually no leak should occur with soft seals on the piston. Replace seals as required.
	2. Other circuit leaks	Check for leaks through the operating valve and correct if necessary. Correct leaks in connecting lines.
Cylinder fails to move the load when valve is actuated	1. Pressure is too low	Check pressure at cylinder to make sure it is in accordance with circuit requirements.
	2. Piston seal leak	Operate valve to cycle the cylinder and observe fluid flow at the valve-exhaust ports when cylinder is at the end of the stroke. Replace seals if flow is excessive.
	3. Piston rod broken at piston end	Disassemble and replace piston rod.
Erratic cylinder action	1. Valve sticking or binding	Check for dirt. Check for contamination of oil. Check for air in the system.

(continued)

(continued from previous page)

Trouble	Cause	Remedy
	2. Cylinder sticking or binding	Check for dirt or air leaks as above. Check for misalignment, or defective packings or seals.
	3. Sluggish operation during warm-up period	Viscosity of oil too high, or pour point too high at starting temperature.
	4. Internal leakage in cylinder	Repair or replace seals and packing. Check oil to see that viscosity is not too low.
	5. Air in system	Bleed air from lines and check oil level, oil condition, and for foamy oil.
Cylinder body-seal leak	1. Excessive pressure	Check maximum pressure rating on the cylinder nameplate. Reduce pressure to the rated limits. Replace seal and retorque bolts.
	2. Pinched or extruded seal	Replace cylinder body seal and retorque the bolts.
	3. Seal deterioration— hard, or loss of elasticity.	Usually this is due to exposure to elevated temperature. Replace the seal.
Rod-gland-seal leak	1. Torn or worn seal	Examine the piston rod for dents, nicks, gouges, or score marks. Replace piston rod if the surface is

Trouble	Cause	Remedy
	2. Seal deterioration	rough. Check the gland bearing for wear. If clearance is excessive, replace gland and seals. Repeat the cylinder body-seal-leak text procedure.
Excessive or rapid piston-seal wear	1. Excessive back pressure due to over-adjustment of speed-control valves (meter-out types)	Correct the valve adjustment.
	2. Wear due to con-taminated fluid	Check for oil con-tamination and clean the fluid, filters, etc. if necessary.

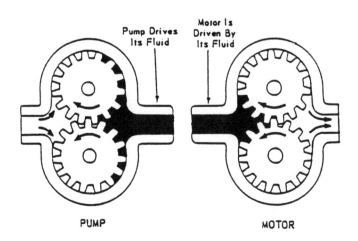

Figure 2.26. Hydraulic pump and motor compared.

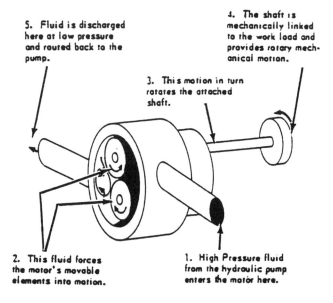

Figure 2.27. Basic operation of a hydraulic motor.

(text continued from page 96)

If we know the displacement and the desired speed of a motor, we find the delivery requirements like this:

$$\text{Delivery (gpm)} = \frac{\text{speed (rpm)} \times \text{displacement (in.}^3\text{/rev)}}{231}$$

For instance, a motor with a displacement of 2.31 in.³/rev, to run at 1,000 rpm, would require a supply of 10 gpm.

$$\text{Delivery (gpm)} = \frac{1,000 \text{ rpm} \times 2.31 \text{ in.}^3\text{/rev}}{231} = 10$$

If we know the displacement and the delivery that is supplied, we can calculate the drive speed in revolutions per minute (rpm).

$$\text{Speed (rpm)} = \frac{\text{delivery (gpm)} \times 231}{\text{displacement (in.}^3\text{/rev)}}$$

From this it can be seen that increasing the displacement of a motor reduces its speed; decreasing the displacement increases speed. The other way to increase speed is to increase the delivery (gpm).

Torque by definition is a turning or twisting effort, or a rotary thrust. Motor torque is usually measured in lb-in. or lb-ft. A torque wrench is simply a wrench with a scale on it to indicate how much twist the wrench is exerting.

The torque a motor develops depends on the load and the radial distance from the center of the motor shaft. For instance, the 5-lb force exerted by the weight in Figure 2.28 is effective five inches from the shaft center. The torque on the motor shaft is equal to the pulley radius multiplied by the weight, or 25 lb-in. In view B, the pulley is smaller, so the torque is less. The small pulley imposes less torque on the motor, but the large pulley raises or lowers the weight faster if the motor speed does not change.

Types of Hydraulic Motors

Motors are designed in three basic types; the same as for the pump: *gear motors, vane motors,* and *piston motors.*

All three designs work on the principle that a rotating unit inside the motor is moved by the incoming fluid.

Gear Motors

Gear motors (Figure 2.29) are widely used because they are simple and economical. Often they are used to drive small equipment in remote applications. Usually small in size, gear motors are versatile and can be transferred from one use to another by using a universal mounting bracket and flexible hoses. Gear motors can rotate in either direction but are not capable of variable displacement.

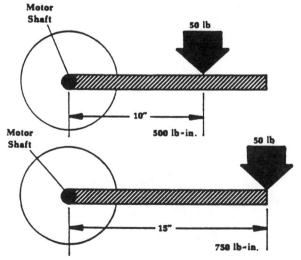

Figure 2.28. Hydraulic motor torque.

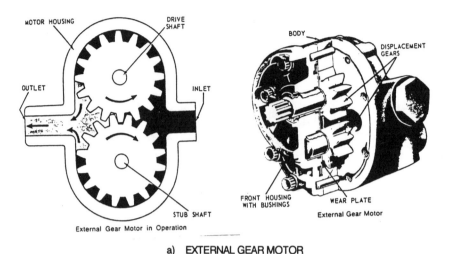

a) EXTERNAL GEAR MOTOR

b) INTERNAL GEAR MOTOR

Figure 2.29. Hydraulic gear motors.

Vane Motors

Vane motors (Figure 2.30) are available in two types—balanced and unbalanced. Most of the vane motors on today's machines are the balanced type, since most of these applications do not require variable displacement. Balanced motors have a longer service life (less bearing wear) and so are more economical.

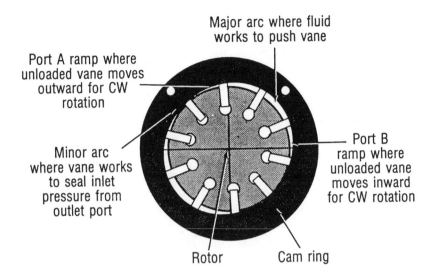

Major arc where fluid
works to push vane

Port A ramp where
unloaded vane moves
outward for CW
rotation

Port B
ramp where
unloaded vane
moves inward
for CW rotation

Minor arc
where vane works
to seal inlet
pressure from
outlet port

Rotor   Cam ring

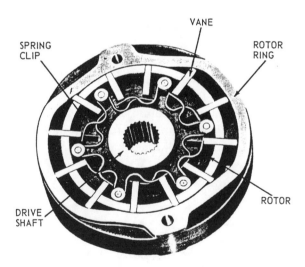

VANE

SPRING
CLIP

ROTOR
RING

DRIVE
SHAFT

ROTOR

Figure 2.30. Hydraulic vane motor.

The balanced vane motor operates much the same as the vane pump. The slotted rotor turns, driven in this case by the force of incoming oil against the vanes.

The only extra feature of the motor is the device used to hold the vanes in contact with the outer ring. These devices may be spring clips, or small springs beneath each vane, pushing them out.

These devices are needed for internal sealing in a motor but not in a pump. In the pump, centrifugal force throws the vanes out against the outer ring. But in the motor, incoming oil is under high pressure and would bypass the vanes before rotation began unless the vanes were held out solid against the ring.

Piston Motors

Piston motors are often favored in systems that have high speeds or high pressure. While more sophisticated than the other two types, piston motors are more complex, more expensive, and require more careful maintenance.

Like its pump counterpart, the piston motor is available in two types: *axial piston* and *radial piston.*

On mobile systems, axial-piston models are often favored. The radial-piston model is usually confined to stationary industrial uses where space is not limited and more power is needed. Figure 2.31 depicts an axial-piston motor.

In operation, high-pressure oil enters the cylinder bores, forcing the pistons against the angled swashplate. Because the swashplate is fixed, the pistons slide down its angled face. This sliding action causes turning of the cylinder block, which in turn drives a shaft that propels the load. As the cylinder block turns, other bores align with the inlet port and their pistons are actuated, keeping up the rotation.

During the second half of the motor's revolution, low-pressure oil is discharged as the pistons are forced back by the thicker part of the swashplate.

To reverse the rotation, the flow of oil into and out of the motor is reversed.

Hydraulic Motor Malfunctions

The great similarity between the motor and the pump was shown in the preceding paragraphs. Unfortunately, this similarity extends into the realm of motor failures as well.

The majority of motor problems fall into these categories:

- Improper fluid
- Poor maintenance or installation
- Improper operation
- Poor motor selection
- Improper system design
- Mechanical failures

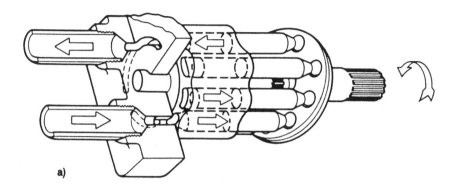

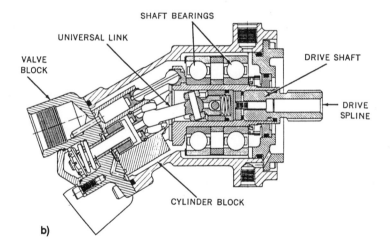

Figure 2.31.   (a) Axial-piston motor. (b) Piston motor in operation.

Improper Fluid

The motor is no different than any of the other components of the hydraulic system—it must have clean fluid, in adequate supply and of the proper quality and viscosity.

Poor Maintenance or Installation

Poor maintenance and installation programs run a close second in the list of major problems. We find many causes of failures in this category, all of which can be attributed to human error. Some of these causes are:

1. Failure to check and repair lines and connections for air or fluid leaks. This can allow dirt and air into the system, plus lower pressure and cause erratic operation.
2. Failure to check and repair other components such as pumps, control valves, or filters. This can lead to excessive wear of internal parts due to excessively high or low pressure, or excessive or inadequate fluid.
3. Failure to install the motor correctly. Motor shaft misalignment can cause bearing wear, which can lead to loss of efficiency. A misaligned shaft can also reduce torque, increase friction drag and heating, and result in shaft failure.

Improper Operation

Exceeding the operating limits is a prime way to promote motor failure. Every motor has certain design limitations on pressure, speed, torque, displacement, load, and temperature.

The list below describes what can happen if these limits are exceeded:

- *Excessive pressure:* Can cause wear due to lack of lubrication, generate heat due to motor slippage, or cause motor to exceed torque limits.
- *Excessive speed:* Can cause heating due to slippage, or cause wear of bearings and internal parts.
- *Excessive torque:* Can cause fatigue and stress to bearings and shaft, especially on applications that require frequent motor reversing.
- *Excessive displacement:* Can generate heat due to pressure across motor not producing usable work.
- *Excessive load:* Can create bearing and shaft fatigue.
- *Excessive temperature:* Can cause loss of efficiency and speed due to thinned oil, and can produce rapid wear due to lack of lubrication.

Diagnosing Motor Failures

Table 2.6 is presented as a general guide to some common motor failures, what the causes might be, and possible remedies.

Table 2.6
Diagnosing Motor Problems

Trouble	Cause	Remedy
Motor won't turn	1. Shaft seizure due to: a. Excessive loads b. Lack of lubrication c. Misalignment 2. Broken shaft 3. No incoming pressure 4. Contaminated fluid	a. Check load and load capacity of the motor. b. Check fluid level and quality. Check operating pressure and temperature. c. Correctly align shaft with work load. Replace shaft. Find reason for breakage. Check and repair clogged, leaking, or broken lines or passages. Check and clean entire hydraulic system. Find source of contamination. Install clean fluid of proper quantity and quality.
Slow motor operation	1. Wrong fluid viscosity 2. Worn pump or motor 3. High fluid temperature 4. Plugged filter	Install clean fluid of proper quantity and quality. Check pump and motor specifications. Repair or replace as necessary. Check for line restrictions, wrong fluid viscosity, or low fluid level. Check cause of plugging, and clean or replace filter.
Erratic motor operation	1. Low pressure	Check for air or fluid leaks.

(continued)

(continued from previous page)

Trouble	Cause	Remedy
	2. Inadequate fluid flow	Check for air or fluid leaks. Check system capacity.
	3. System controls failing	Check pump and control valves for proper operation.
Motor turns in wrong direction	1. Pump-to-motor connections wrong	Reconnect pump and motor.
	2. Wrong timing	Check specifications.
Motor shaft not turning	1. Excessive work loads	Check motor load specifications.
	2. Internal motor parts worn or broken	Replace parts or complete motor as necessary.
Motor not turning over or not developing proper speed or torque	1. Drive mechanism binding because of misalignment	Remove motor and check torque requirement of driven shaft.
	2. Free recirculation of oil to reservoir	Check circuit, valving, and valve position.
	3. Sticky relief valve (open)	Remove dirt from under pressure-adjustment ball or piston.
	4. Setting of overload valve not high enough	Check system pressure and reset relief valve.
	5. Pump not delivering sufficient oil	Check pump delivery, pressure, and motor speed.
External oil leakage	1. Gaskets leaking (may be due to reservoir drain not being connected if this is required)	Replace. (If drain line required it must be piped directly to reservoir with little back pressure.)

Accumulators

Accumulator Types and Applications

An accumulator is a storage device for high-pressure hydraulic oil. The system pump stores oil in the accumulator during periods when it would normally be unloaded or idling. The stored oil is available at a later time, either to supplement pump oil, or for use when the pump is shut down.

One very common type of accumulator has an appearance similar to a fluid-power cylinder, with the piston acting as a barrier between the stored high-pressure oil and the compressed gas, which is the energy-storing medium. Fluid pressure on both sides of the piston is essentially the same. Volume rating of an accumulator (1 gal, 2 gal, 5 gal, etc.) is the approximate cubic capacity of the gas side when all oil is discharged. Volume is slightly less on the oil side due to the piston shape.

Besides the piston-type accumulator, several other configurations are available, as shown on Figure 2.32. Included among these are the following:

- *Spring-loaded piston type:* Where a spring is used to preload the piston instead of a gas.
- *Diaphragm and bladder types:* In these units, an elastomeric (rubber) member is used to separate the gas from the oil. In the bladder type, the gas is inside and the oil surrounds the rubber bladder. In these units, as the oil leaves the accumulator, the rubber member expands and eventually conforms to the internal shape of the accumulator when it is empty of oil.
- *Weighted type:* Physical weight responding to gravity forces is supported by a hydraulic cylinder. The area of the piston in square inches and the mass being supported determines the resulting pressure in lb/in.^{2}. Frictional forces are normally negligible. Obviously, the weighted accumulator must be installed in a vertical mode. The weighted-type accumulator provides a uniform pressure, from a completely-charged condition to minimum-charged condition. The weight and area relationship is a constant value. Consequently, the weighted-type accumulator is well qualified to provide uniform force for holding various types of forming rolls in position and to maintain pressure during long holding cycles, such as rubber- and plastic-molding activities. The weighted accumulator is provided

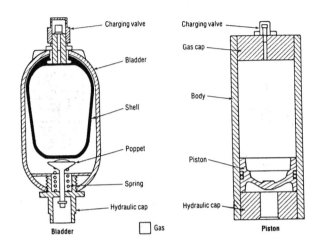

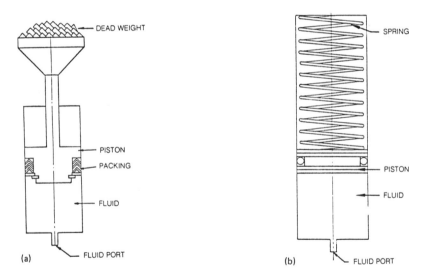

Figure 2.32.   Accumulator types.

with mechanical signal sources to control charge function. The spring-loaded and gas-charged accumulators depend on a pressure signal to control the charge function.

A piston accumulator may be mounted in any position, although the preferred arrangement, where there will be a very heavy momentary discharge, is to position it so that the heavy discharge is in a straight line with the work. Potential contamination in the hydraulic fluid may make it desirable to mount the piston-type accumulator vertically to avoid scratching the barrel by solid particles lodged in the piston seals. The preferred mounting position for bag-type accumulators is vertically upright, although they will operate in any position. Contamination in a bag-type accumulator can become embedded in the bladder, creating a vulnerable area, or prevent the proper operation of the anti-extrusion valve, allowing the bag to partially extrude into the oil area when pressure in the hydraulic fluid decreases.

Accumulators can be considered energy-storage devices similar to a receiver tank in a pneumatic system. However, they can be used for other purposes, as shown in Figures 2.33 and 2.34, including as shock absorbers (Figure 2.35), as pressure surge dampeners, and for reservoir pressurization where it is undesirable for the tank to be open to the atmosphere through a breather.

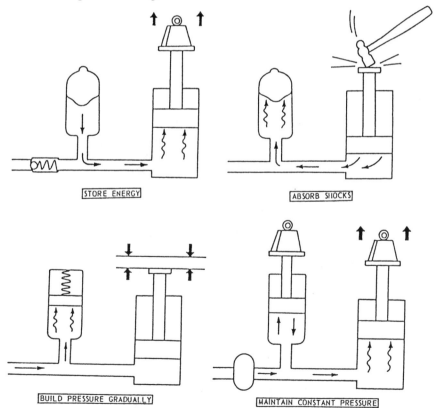

Figure 2.33.   Accumulator applications.

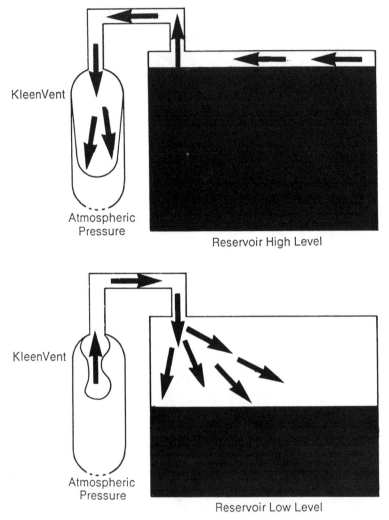

Figure 2.34. Closed reservoir arrangement using bladder accumulator.

Bladder-type devices provide an excellent barrier, preventing dirt and moisture from entering hydraulic reservoir. The reservoir can be completely sealed, and the differential area created by the cylinder rods can be accommodated by the captive air in the reservoir plus the excursion of the bladder in the barrier device. One side of the barrier device is to atmosphere and the other side is to the captive air on top of the oil in the reservoir. Air pressure on the reservoir may range up to several psi, depending on the location of the reservoir, type of

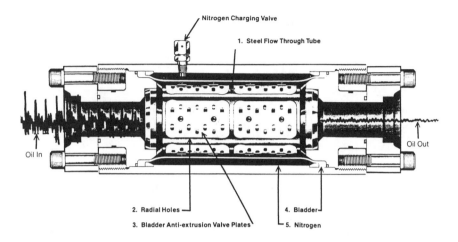

Figure 2.35. Hydraulic shock absorber.

construction, and design of the barrier assembly. Some hydraulic reservoirs are designed to operate with positive pressure on the captive air above the oil. This positive pressure may be used to provide better pump intake conditions.

Servicing and Repair

Observe the following precautions when working on pneumatic accumulators.

1. *Caution: Never fill an accumulator with oxygen!* An explosion could result if oil and oxygen mix under pressure.
2. Never fill an accumulator with air. When air is compressed, water vapor in the air condenses and can cause rust. This in turn may damage seals and ruin the accumulator. Also, once air leaks into the oil, the oil becomes oxidized and breaks down.
3. Always fill an accumulator with an inert gas such as dry nitrogen. This gas is free of both water vapor and oxygen; this makes it harmless to parts and safe to use.
4. Never charge an accumulator to a pressure more than that recommended by the manufacturer. Read the label and observe the "working pressure."
5. Before removing an accumulator from a hydraulic system, release all hydraulic pressures.

6. Before you disassemble an accumulator, release both gas and hydraulic pressures.
7. When you disassemble an accumulator, make sure that dirt and abrasive material does not enter any of the openings.
8. Install a valve to vent pressurized oil from accumulator to tank, preferably automatically when machine shuts down. This may be mandatory in some locations as a part of a safety code.

Servicing and Precharging Pneumatic Accumulators

Checking a precharged accumulator on the machine.

1. If you suspect external gas leaks, apply soapy water to the gas valve and seams on the tank at the "gas" end. If bubbles form, there is a leak.
2. If you suspect internal leaks, check for foaming oil in the system reservoir and/or no action of the accumulator. These signs usually indicate a faulty bladder or piston seals inside the accumulator.
3. If the accumulator appears to be in good condition but is still slow or inactive, precharge it as necessary.

Before removing an accumulator from a machine. First, be sure that all hydraulic pressure is released. To do this, shut down the pump and cycle some mechanism in the accumulator hydraulic circuit to relieve oil pressure (or open a bleed screw).

Removing an accumulator from a machine. After all hydraulic pressure has been released, remove the accumulator from the machine for service.

Repairing an accumulator.

1. Before dismantling an accumulator, release all gas pressure. Unscrew the gas-valve lever very slowly. Install the charging valve first, if necessary. Never release the gas by depressing the valve core, as the core might be ruptured.
2. Disassemble the accumulator on a clean bench area.
3. Check all parts for leaks or other damage.
4. Plug the openings with plastic plugs or clean towels as soon as parts are removed.
5. Check bladder or piston seals for damage and replace if necessary.
6. If gas-valve cores are replaced, be sure to use the recommended types.
7. Carefully assemble the accumulator.

Charging Accumulators

Energy is stored in the accumulator by compression of an inert gas. Any of the inert gases could be used, but for economic reasons, *nitrogen is normally used because it is nonflammable and is readily available* on the industrial market at a reasonable price. It is obtained in high-pressure bottles of approximately 2,200 psi. Where possible, use oil-pumped "dry" nitrogen to reduce the possibility of moisture condensation in the gas chamber, which would seriously damage a piston-type accumulator.

The actual pressure (psi) value of the nitrogen precharge pressure is not highly critical. A range of one-third to one-half of the maximum hydraulic pressure will give good results on most applications. Very little difference in circuit performance will be noticed within this range. The suggested procedure is to precharge initially to the higher value, then do not add more gas until the pressure falls to the lower value. The effect of different precharge pressures is shown on Figure 2.36.

> *Caution:* Do not use oxygen or any gas mixture containing oxygen, such as compressed air, for precharging accumulators that have rubber seals or separators. Oxygen deteriorates rubber by oxidation and the use of oxygen always creates a possible fire or explosion hazard in any situation. This is especially dangerous in the vicinity of hydrocarbons, such as petroleum oil.

One way to test for insufficient accumulator-oil capacity is to install two pressure gauges, one on the oil port, the other on the gas port. Observe these two gauges while the machine is going through its cycle. They should have essentially identical readings throughout the entire cycle unless the accumulator runs out of oil. In this case, the oil gauge will suddenly drop to near zero as the piston in the accumulator bottoms out against the end cap. If the accumulator runs out of gas, or the precharge drops too low, the machine will be slow throughout the cycle, working only at the speed from oil flow developed by the pump alone.

Precharging equipment. A charging-and-gauging hose assembly (Figure 2.37) should be purchased from the manufacturer of the accumulator brand in use, and should be kept on hand as a service tool. Because of variation in construction of the air valve, a charging hose for one brand may or may not work on a different brand.

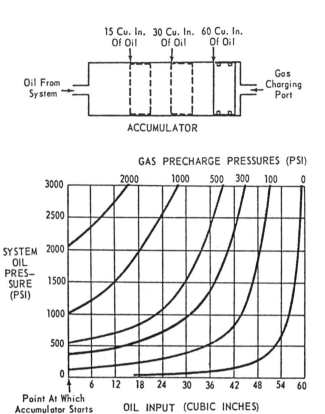

Figure 2.36.   Effect of different precharges on a pneumatic accumulator.

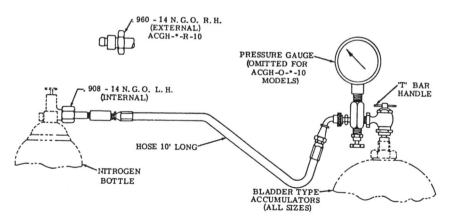

Figure 2.37.   Charging and gauging assembly.

Procedure for precharging. The accumulator may be precharged either while connected into the system or as a loose component lying on the test bench. First, make sure that all oil is discharged from the accumulator. If it is connected into a hydraulic system, open the circuit bleed-down valve to discharge all accumulator oil to the reservoir. Next, connect the charging hose to the accumulator and take a reading of the existing precharge pressure, if any. Connect the hose to the gas bottle. A pressure regulator need not be used. It is not difficult to control the flow rate of gas with the shutoff valve located on the bottle. Slowly open the valving and allow gas to flow into the accumulator. The precharge level cannot be accurately gauged while gas is flowing. Shut off the valve frequently to observe the psi level in the accumulator. After precharging, immediately close all valves and remove the charging hose. For safety, replace the cover on the gas bottle and the cover on the gas valve of the accumulator.

Frequency of Servicing

The gas precharge will gradually leak down on any accumulator. The gas molecules transfer through the pores of the rubber by osmosis. Piston-type accumulators should be tested every month or two, more often if found to be necessary. On rapid-cycling applications, precharge psi should be checked every 100,000 cycles. Bag or bladder accumulators have a much greater leak-off rate because of the much greater rubber surface, and should be serviced more often. With piston accumulators, the piston seals (O-rings) should be replaced approximately every 1,000,000 cycles because of mechanical wear. Some accumulators use O-rings of special cross section, and although it is possible to replace these with standard commercial rings, it is better to obtain exact replacements from the manufacturer.

Rule of thumb. For every 1% an accumulator is oil charged above the minimum system pressure, it will deliver 1 in.^{3} of oil for each gallon of its rated size. *Example:* Assume a hydraulic system where the accumulators are to be charged to a maximum pressure of 3,000 psi. System pressure will be permitted to fall to 2,000 psi after the accumulator has delivered its oil. From 2,000 to 3,000 psi represents an increase of 50% that the accumulator will be charged above minimum pressure. According to the rule of thumb, a 1-gal accumulator would give out 50 in.^{3} of oil under these conditions. If the circuit required 250 in.^{3} of oil, you would select a 5-gal accumulator.

Installing an accumulator on a machine. Attach the accumulator to the machine and connect all lines. Start the machine and cycle a hydraulic function to bleed any air from the system. Then check the accumulator for proper action.

Details of accumulator troubleshooting are given in Table 2.7. Some examples of accumulator damage are shown on Figure 2.38.

Hydraulic Reservoirs

Although a hydraulic reservoir has no moving parts, it is an important component of its hydraulic system. It must supply, purify, and cool the fluid. Hydraulic reservoirs are built in a variety of shapes and sizes to fit the requirements of the system involved. A satisfactory reservoir must meet the following requirements (Figure 2.39):

- It must be leakproof. Leaks are costly, and may cause fires and falls.
- It must be rustproof. If a petroleum-based fluid is used, the interior of the reservoir should be painted. Many synthetic fluids will protect the interior of the reservoir adequately without painting; if painting is required, the paint must be compatible with the fluid.

Table 2.7
Diagnosing Accumulator Problems

Problem	Cause	Remedy
Slow reaction	1. Loss of charge or overcharge	Check charge pressure; reset
	2. Unloading valve or pump pressure set too low	Adjust to higher pressure
	3. Relief valve set too low or stuck open	Reset or clean valve
	4. Pump not pumping	Check pump
	5. Unloading pressure switch set too low	Reset pressure switch
Fails to absorb shock	1. Loss of charge or overcharge	Check and recharge if necessary or correct charge pressure

a)   LOSS OF BLADDER ELASTICITY

b)   BLADDER CUT FROM POPPET

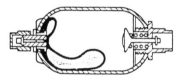

c)   EXTRUSION INTO GAS VALVE (NO PRE-CHARGE)

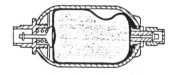

d)   UNEVEN BLADDER WEAR FROM HORIZONTAL
     ACCUMULATOR MOUNTING

Figure 2.38.   Accumulator damage examples.

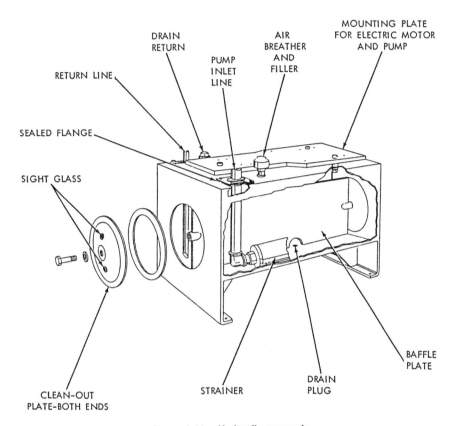

Figure 2.39. Hydraulic reservoir.

- It must hold enough fluid to meet the maximum requirements of the system without exposing the suction strainer. When large-bore, long-stroke cylinders with oversized rods are employed, this requirement may be critical. For example, a 16-in. bore cylinder with a 100-in. stroke and a 12-in.-diameter rod will use about 87 gal of fluid on the forward stroke, but only about 38 gal on the return stroke. This difference means that, on the forward stroke, 49 gal will be drawn from the reservoir.
- It must have sufficient capacity to control the fluid's operating temperature, unless an external air- or water-cooled heat exchanger is employed. Even a well-designed hydraulic system will convert 20% of its input power to heat; this heat must be removed from the fluid.

- It must have sufficient capacity to separate water dispersed in a petroleum-based fluid.
- It must have sufficient capacity to allow solid contaminants to settle to the bottom. (Baffles separating the intake from the return can be of considerable help in promoting settling.)
- It must have sufficient capacity to allow trapped air to escape. Air bubbles entering the intake can cause pump cavitation and jerky cylinder action. Baffles and antifoaming additives assist the reservoir in performing this function.
- It must be easy to clean. A drain opening should be located so the fluid can be emptied before the cleanout plates are removed. Ample access to the interior should be provided.
- It must be sturdy enough to support the components that will be mounted on it—often the pump, pump motor, motor coupling, relief valve, and some control valves. Sometimes a mounting plate is attached to the top of the reservoir to hold this equipment. In other instances, it is mounted directly on the reservoir top. In either case, the top, sides, and bottom of the reservoir must be stiff enough that they will not act as resonators and amplify the pump's noise.

Design

The bottom of a reservoir should be several inches above the floor. Old designs were often set directly on the floor. Not only did this make it impossible to clean up spilled fluid that had seeped under the reservoir, but it cost the reservoir about 25% of its cooling area.

Baffles (Figure 2.40) should be provided in the reservoir to keep the incoming fluid separated from the outgoing fluid for as long as possible. These are always vertical, but may run in either direction, depending on the shape of the reservoir.

Most industrial hydraulic reservoirs are rectangular. Low, flat tanks are seldom satisfactory—in general, the height of the reservoir should be at least equal to its smallest horizontal dimension.

L-shaped reservoirs are also popular and practical. In this design, the pump, motor, relief valve, and heat exchanger are mounted on the base of the L, and the control valves are on the sides of the vertical section above the fluid level. If subplate valves are used, nearly all the piping will be inside the reservoir. The drain lines will be above the level of the fluid, so no back pressure will be put on the valves. This placement also permits a visual check of the amount of fluid expelled by the drain lines while the circuits are in operation.

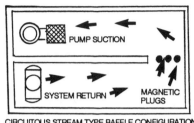

CIRCUITOUS STREAM TYPE BAFFLE CONFIGURATION

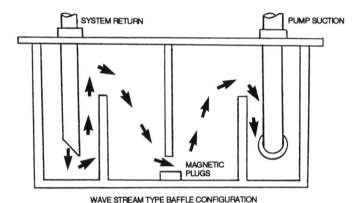

WAVE STREAM TYPE BAFFLE CONFIGURATION

Figure 2.40.   Baffle arrangements.

The L-shaped design provides large surface areas for cooling. The bottom of the reservoir should be about 6 in. above the floor to allow room for air circulation.

T-shaped reservoirs actually take the form of inverted T's. The extra leg on the bottom permits the reservoir to carry an extra pump with its associated equipment. Both pumps use the common reservoir.

In some power-unit designs, the electric motor and the pump are mounted vertically, with the pump inside the rectangular reservoir. The pump may or may not be submerged in the fluid; if it is submerged, problems with suction-side air leaks are eliminated.

On some large rectangular reservoirs, the control console panels are welded or bolted to the reservoir. This design keeps the controls off the machine and permits easy servicing. Subplate or manifold-mounted valves are used, so that the only piping between the reservoir unit and the machine are the lines to the actuators.

Sizing

Location, operating temperature, operating cycle, and piping configuration will all affect reservoir size. For example, if a relief valve allows fluid to spill for long periods during the cycle, a great deal of heat will be generated; unless the system is equipped with a heat exchanger, a reservoir large enough to dissipate this heat will be required. If the reservoir is to be located where the ambient temperature is high, a large reservoir may be needed.

An often-quoted rule of thumb is that the capacity of the reservoir in gallons should be two to three times the capacity of the pump in gpm—variations in system design can make this rule very inaccurate. Some highly engineered systems have reservoirs with half the pump's capacity. Manufacturers of "packaged" hydraulic power units (in which the reservoir and the pumps are parts of the package) offer several pump sizes for each reservoir size. For example, with a 20-gal reservoir, the buyer can choose among 3-, 4.5-, 5.5-, 7.5-, and 8.5-gpm pumps; with a 70-gal reservoir, the buyer may select a 20- or a 30-gpm pump.

Some synthetic fluids require large reservoirs for foam separation. Also, some systems need large intake filters—these may dictate the size of the reservoir.

Auxiliary Equipment

Reservoirs should incorporate a level gauge marked with the permissible high and low fluid levels. They should also have temperature gauges. (Devices are available that combine both instruments.)

Oil-filler assemblies should be of sufficient size to allow filling without a spill. They should be equipped with screens that will prevent particles larger than 40 microns from entering the reservoir.

Unpressurized reservoirs require an air-breather assembly. These normally incorporate a filter that provides about 40-micron filtration. In extremely dirty locations, it may be necessary to install an oversized breather to prevent the filter from becoming clogged. The breather assembly should never be removed except for cleaning, and the filter should be checked regularly.

Sometimes it is advisable to install a magnetic tank cleaner in the reservoir to remove fine metallic particles. These cleaners will pick up ferrous particles with weights as small as 1 grain (0.002 oz). It is estimated that in some instances the life of a pump can be doubled by the use of a magnetic tank cleaner; but, like filters, these cleaners require servicing at regular intervals.

Location

The siting of the reservoir can affect the performance of the entire hydraulic system; it should be considered carefully. Here are some do's and don'ts:

- Don't put the reservoir in a pit. Ventilation will be poor, and heat buildup will cause system malfunctions. Pits also make maintenance difficult.
- Don't put the reservoir against a wall, since air circulation will be restricted.
- Do keep the reservoir accessible. Don't allow rubbish, boxes, or pallets to accumulate around it.
- Don't put a sling or lifting hook around the pump shaft or the motor shaft when moving the reservoir into position. Most reservoirs have lifting eyes or lifting holes, or an area in which the forks of a lift truck can be inserted.
- Don't forget that a full reservoir is heavy. A 20-gal power unit can weigh well over 300 lb.
- Don't forget to bolt the reservoir down. The feet of most units are furnished with mounting holes.
- Don't forget to protect the reservoir from blast furnaces, heat-treating equipment, and rolling mills. Plenty of heat is created in hydraulic systems without adding more from external sources.
- Do keep the reservoir close to the machine it serves—proximity reduces the time required for cylinders to act.
- Don't install a reservoir on a lower floor to operate a machine on the floor above. There have been instances in which the amount of fluid in the lines caused a de-energized pump to run backwards, and the fluid overflowed the reservoir. Even with check valves, this type of installation is likely to give trouble.
- Do keep the fluid level in the reservoir below the controls.
- Don't expose the reservoir to low ambient temperatures if it can be prevented.
- Do install thermostatically controlled heaters, if a cold location cannot be avoided.

Plant engineers will occasionally encounter reservoirs built into the base or head of a machine. A press reservoir is often built into the crown-head of the press, with the controls mounted on the reservoir and the control actuators on a panel on the press. The advantage of this design is that it conserves floor space. Its disadvantages are that it is difficult to add fluid, clean the reservoir, and repair the controls, and that back pressure on the valve drain lines may be a problem.

Cleaning

A reservoir may be emptied by attaching a pipe or hose to the drain outlet (which should be equipped with a shut-off valve), or by removing the screen in the filler assembly and pumping the fluid out through a hose. When the reservoir is empty, the clean-out plates should be removed and the interior of the reservoir wiped out with lint-free rags. Rust should be removed. If the system uses a petroleum-based fluid, the interior should be repainted, unless the old paint is still in good condition. The air-breather assembly and the filler assembly should be cleaned and replaced. Before the clean-out plates are reassembled, the condition of their gaskets should be checked; a faulty gasket will cause a continuous seepage of fluid until the reservoir is emptied again so that the gasket can be changed. After the shut-off valve has been closed and the pipe or hose connection to the drain line removed, the reservoir should be filled to the high-level mark on the sight gauge with an approved fluid.

Cleaning intervals will depend on the severity of service, the amount of contamination encountered, the quality and type of fluid used, the ambient temperature, and various other conditions. Fluid quality should be checked regularly, and the reservoir cleaned whenever it is necessary to change or reprocess the fluid.

Hydraulic Fluids

Leading hydraulics-system designers regard hydraulic fluids as the single most important group of materials in hydraulic systems. The fluid is literally the lifeblood of the system, the one element that ties everything together. With proper selection and handling of the fluid, most of the potential problems with the system can be prevented and the system can function more nearly as it was designed.

Mistakes and oversights in selecting, storing, and installing the fluid or in maintaining the hydraulic system can cause no end of trouble for operators and maintenance people. According to fluid-power industry spokespeople, between 70 and 85% of all hydraulic system problems are directly related to improper choice or handling of hydraulic fluids (Mobil Oil Corp., 4/89).

Heat and contamination degrade fluids, while at the same time they attack other system components—pumps, seals, cylinder walls, etc. The result may take some time to surface, but when it does, you can face costly downtime while the cause is determined and a remedy applied.

People in the industry believe that fully 85% of all hydraulic fluid ever installed leaks out—either slowly, or in major line breaks, or by failures of fittings, seals, and the like. Besides the obvious wastefulness, fluid jets, sprays, and gushers can be very dangerous. Some are capable of penetrating the skin or damaging the eyes and other organs.

One expert from Texaco Corp. estimates that 7 million barrels of hydraulic fluids are lost each year through leaks and line breaks.

Besides causing human hazards and loss of production, fluid leaks are a major cause of fires when ignition sources are nearby. Growing recognition of the serious fire-risk problem has led to the widespread use of fire-resistant fluids in many high-hazard areas.

There is no such thing as a universal or ideal hydraulic fluid. One major reason is the imposing list of fluid characteristics considered important by users, system designers, and manufacturers. For that reason, selection of the proper fluid for a given application is virtually always a compromise.

To do its job well, a hydraulic fluid must do at least six things:

1. Transfer fluid power efficiently
2. Lubricate the moving parts
3. Provide bearings in the clearances between parts
4. Absorb, carry, and transfer heat generated within the system
5. Be compatible with hydraulic components and fluid requirements
6. Remain stable against a wide range of possible physical and chemical changes, both in storage and in use

Resistance to oxidation is particularly significant. Burning, of course, is an oxidation process. Slower oxidation reactions give rise to fluid degradation with resultant formulation of such troublesome reaction products as sludge, varnish, and gum, or with the formation of corrosive fluid that can attack metallic components. Other changes that need to be resisted include physical wear and pitting on pipe, tubing, and component surfaces; excessive swelling or shrinking of seals, gaskets, and other materials; significant viscosity variation; foaming; and evaporation.

Hydraulic fluid may travel through a system at velocities of 15 or 20 ft/sec, or more. In compact, mobile systems with small reservoirs, very high turnover may pass the fluid completely through the circuit two or more times per minute. On the other hand, where the reservoir is large, the fluid may get to "rest" a bit between cycles, allowing it to transfer more of its heat, and release entrained foam-causing air from the system.

Fluid Types

The widespread use of the term "hydraulic oil" reflects the dominance of petroleum-based hydraulic fluids. Hydrocarbon oils protect well against rust, have excellent lubricity, seal well, dissipate heat readily, and are easy to keep clean by filtration or gravity separation of contaminants.

Petroleum Oils

Petroleum oil is a quite serviceable industrial hydraulic fluid when specifically refined and formulated with various additives to prevent rust, oxidation, foaming, wear, and some other problems, so long as heat and fire hazards are not critical. But, with all their useful properties, hydrocarbon oils do have one important drawback—they burn, and at temperatures a good bit lower than a number of other hydraulic-fluid types.

For that reason, several types of fire-resistant hydraulic fluids are available, most of them more costly than petroleum-based fluids, and several of which give away something in performance for the added resistance to burning.

It should be noted that the term "fire-resistant" is far from meaning "fireproof." Under certain conditions, almost any fluid can burn. However, fire-resistant fluids resist ignition, while petroleum oils ignite quickly and propagate a blazing flame.

Fire-Resistant Fluids

Apart from the so-called water-additive hydraulic fluids, there are four principal types of fire-resistant hydraulic fluids. Following are some of their principal advantages and limitations:

1. Phosphate esters. Sometimes called the straight synthetic fluids, the triaryl phosphate esters are excellent lubricants. In fact, they are the best among the fire-resistant (F-R) types. They have good fire resistance and are better at higher temperature ranges and at higher pressures than many other F-R fluids. However, they are less useful at lower temperatures, their high specific gravity requires care in selecting pumps, and they are the most costly of all the industrial fire-resistant hydraulic fluids.

2. Water glycols. These fluids are true solutions, not emulsions, containing a three-component mixture of water (35 to 40%), a glycol, and a high-molecular-weight water-soluble polyglycol. They have excellent fire resistance, good lubricating properties, and are available in a range of viscosities. However, they should not be used at temperatures higher than 120°F, and require periodic checks on water content and additive levels because of evaporation.

3. Water-in-oil emulsions. These fluids, sometimes termed invert emulsions, are intended for moderate-duty F-R applications. They consist of 35 to 40% water dispersed in petroleum oil by means of an emulsifying additive package. They have adequate viscosity for hydraulic service; however, although superior to petroleum oil, water-in-oil emulsions do not have inherent fire-resistance of either phosphate esters or water-glycol fluids. They require greater care to avoid contamination and should not be repeatedly frozen and thawed, which will cause the two fluid phases to separate.

4. Oil/synthetic blends. Where fire hazards are moderate, blends of phosphate esters and refined petroleum stocks are increasingly used, together with a coupling agent to stabilize the solution. They have good lubricating properties. Their fire-resistance reflects their composition, and depends largely on the ratio of phosphate ester to petroleum oil.

Fire-resistant hydraulic fluids need careful selection for compatibility with system components and even more care to keep them contamination-free than petroleum oils. A careful evaluation of the fire hazard is important before using these fluids, both for safety and for economy.

For most users of hydraulic systems, the best rule of fluid selection is to follow the system manufacturer's recommendations. The manufacturer has chosen a fluid type that meets the system's various requirements and demands, and has to live with the results of that choice. Just as it is folly to mix two or more different hydraulic fluids in use, so is it sheer trouble to install a type of fluid other than the one suggested by the hydraulic equipment manufacturer.

5. Water. Water in hydraulic systems has had an odd history. Used in some of the very first hydraulic systems, it was soon replaced by petroleum oils because of the limitations of water—freezing, evaporation, corrosiveness, poor lubricity, and low viscosity. Now, with

rising consciousness of fire risks, plus the concern over petroleum supplies and the high cost of synthetic chemical fluids, water hydraulics
is again attracting widespread interest.

The so-called water-additive fluids are of two basic types: oil-in-
water emulsions "soluble oils" (water is the continuous phase) and,
water plus chemical additives. The oil-in-water emulsions contain 60
to 65% oil dispersed in water, while the chemicals-in-water fluids contain only 2 to 5% chemicals and 98 to 95% water.

Making good use of the positive properties of water, such as its low
cost, ease of disposal, and nonflammability, these substances are looked
upon as the next-generation hydraulic fluids for quite a few industrial
and mining hydraulic systems. Industry personnel report that manufacturers of pumps and other hydraulic system components have developed
appropriate hardware.

Properties of Hydraulic Fluids

Viscocity

For proper power transmission, this is a most important property.
Viscosity is a measurement of a fluid's resistance to flow. Said another
way, it is a fluid's "thickness" at a given temperature. Viscosity is
expressed by SAE (Society of Automotive Engineers) numbers 5W,
10W, 20W, 30W, 40W, etc. All petroleum oils tend to become thin
as the temperature goes up, and to thicken as the temperature goes
down. If viscosity is too low (fluid too thin), the possibility of leakage
past seals and from joints is increased. This is particularly true in pumps,
valves, and motors, which depend on close-fitting parts for creating
and maintaining proper oil pressure. If viscosity is too high (fluid too
thick), sluggish operation results and extra horsepower is required to
push the fluid through the system. Viscosity also has a definite influence
on a fluid's ability to lubricate moving parts.

Viscosity is determined by measuring the time required for 60 cubic
centimeters (cm³) of an oil at a temperature of 210°F to flow through
a small orifice in an instrument known as a Saybolt Viscometer or
another instrument called a Kimematic Viscometer. The actual SAE
number is determined by comparing the time required for the oil to
pass through the instruments with a chart provided by the Society of
Automotive Engineers.

Viscosities of common hydraulic fluids are presented in Table 2.8,
and typical recommended viscosities for various pump types are given
in Table 2.9.

Table 2.8
Viscosities of Common Hydraulic Fluids

Fluid	Viscosity (SSU 100°F)
Plain water	30
High-water synthetics	30
Water glycol	200
Phosphate ester	230
Oil-synthetic blend	300
Water-in-oil emulsion	450
Petroleum oil	215

Table 2.9
Typical Recommended Viscosities

Component	Viscosity Range (SSU)	
	Permissible	Optimum
Vane pump, 1,200 rpm	80–1,000	125–250
Vane pump, 1,800 rpm	100–1,000	120–250
Vane motor	80–1,000	120–250
Radial-piston pump	60–300	80–220
Axial-piston pump or motor	40–350	80–200
Gear pump (industrial applications)	40–1,000	120–250

Viscosity Index (VI) is a measure of a fluid's change in thickness with respect to temperature. Fig. 2.41 shows the behavior of a variety of fluids with variation in temperature.

If a fluid becomes thick at low temperatures and very thin at high temperatures, it has a *low* VI. On the other hand if viscosity remains relatively the same at varying temperatures, the fluid has a *high* VI. As pointed out earlier, in a fluid with good viscosity characteristics, there is a balance between a fluid thick enough to prevent leakage and provide good lubrication while at the same time being thin enough to flow readily through the system. Therefore, a fluid with a high VI is almost always desirable.

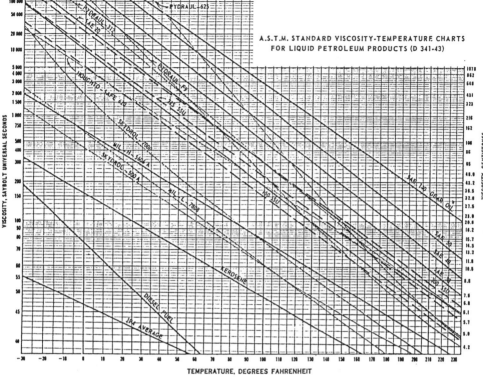

Figure 2.41. ASTM standard viscosity-temperature charts for liquid petroleum products.

Even though carefully refined oils have a good VI, a substance called a viscosity index improver is often added to hydraulic fluids. This substance increases the VI of the fluid so that the change in its viscosity over a wide range of temperatures is as little as is practical.

Wear Prevention

Most hydraulic components are fitted with great precision, yet they must be lubricated to prevent wear. To provide this lubrication, a satisfactory fluid must have good "oiliness" to get in the tiny spaces available between moving parts and hold friction between the parts to a minimum.

There are many other properties or characteristics that a good hydraulic fluid must have. These are listed in Table 2.10.

Maintaining Fluid Performance

Fluid filtration is very important to keep contamination levels down. Many of the newer fire-resistant fluids require even more attention

Table 2.10
Characteristics of a Good Hydraulic Fluid

Characteristic	Definition/Comment
Favorable viscosity	Thickness or resistance to flow; needs to be suited to requirements of the system.
Viscosity-temperature relationship	Variation of viscosity with temperature, described often as viscosity index; minimal change of viscosity with change in temperature is desired.
Chemical and environmental stability	Minimal change in storage or use
Good lubricity	Film formation and friction reduction.
Compatibility with materials	Minimal effect, physically or chemically, on seals, gaskets, hoses, etc., as well as on metallic components.
Heat-transfer capability	High specific heat and thermal conductivity to carry and dissipate heat.
High-bulk modules	Stiffness or low compressibility of the fluid; stiff fluids permit stable handling of heavy loads without "sponginess."
Low volatility	Minimal evaporation and bubble formation reduces fluid loss and dangerous cavitation.
Low foaming tendencies	Foam or entrained air reduces fluid stiffness, another cause of "sponginess."
Nontoxic and nonallergenic properties	To meet various regulatory and safety standards affecting workers, as well as acceptable odor.

to filtration than do conventional hydraulic oils. Selection of appropriate filters, placement in the proper parts of the system, and careful maintenance are all vital.

Checking the quantity of fluid in the system is important. Insufficient fluid can limit complete extension of the cylinders. Low fluid levels can also draw air into the system, creating "spongy" cylinder action and possibly setting up the conditions for one of the more costly hydraulic system problems—pump cavitation and the resultant high or catastrohpic wear.

Cavitation is a compression/expansion process in which tiny gas or vapor bubbles expand explosively at the pump outlet, causing metal erosion, leading ultimately to pump destruction. Some experts regard the action as implosion, rather than explosion. Be that as it may, what the user will hear is a buzzing, rattling sound. This is an early warning signal of cavitation. The sound is as though a handful of stones were tossed into the pump.

When it comes to checking the fluid quality, operators should periodically inspect the cleanliness, color, thickness, or viscosity, and perhaps the odor of the fluid. Beyond these checks, there are numerous standard laboratory tests that can be used to determine everything from foaming tendencies and load-carrying ability to oxidation and thermal stability.

The key to checking fluids in service is to look for changes in fluid properties. Such changes represent warning signals that can indicate a need for corrective action.

Two other points bear repetition in any discussion about the proper maintenance of fluid in a hydraulic system. First, with the problems and costs associated with fluid leakage, good practice must include thorough attention to system integrity of leak-tightness to minimize this difficulty. In some instances, operators have gone so far as to weld many or most connections, although this is often impractical.

Second, and bearing on the destructive effects of excessive heat on fluids and other system components, good operating practice limits the temperature range within which a hydraulic system operates. Proper design and maintenance of the fluid reservoir is important in this respect. Heat exchangers are commonly used in hydraulic systems to aid heat dissipation.

In order to assure proper system operation, maximize fluid life, and minimize fluid consumption, the following suggestions or pointers should be heeded:

1. Use only the type and grade hydraulic fluid recommended; do not mix fluids.
2. Store fluid containers inside or under a roof and on their sides to minimize entry of water and dirt.
3. Clean the cap and the drum top thoroughly before opening.
4. Use only clean hoses and containers to transfer fluid from cans or drums to the hydraulic reservoir; use a fluid-transfer pump equipped with a 25-micron filter.

5. Use a screen on the reservoir filler pipe.
6. Flush and refill the system exactly as recommended; be certain to fill to the proper level, overfilling is as troublesome as underfilling.
7. Check the hydraulic system thoroughly to eliminate leaks or contaminant entry points; keeping hydraulic fluid out of plant effluent makes more sense than having to separate it later for proper disposal.
8. Never return leaked fluid to the system.

Pneumatic Components

Pneumatic Valves

The basic pneumatic valve is a mechanical device consisting of a body and a moving part that connects and disconnects passages within the body. The flow passages in pneumatic valves carry air. The action of the moving part may control system pressure, direction of flow, and rate of flow.

Valve Types

Control of Pressure

Pressure in a pneumatic system must be controlled at two points: at the compressor and after the air-receiver tank. Control of pressure is required at the compressor as a safety measure for the system. Control of pressure at the point of air usage provides a means of safeguarding the system and of maximizing the efficiency—compressed air will not be wasted. Characteristically in a pneumatic system, energy delivered by a compressor is not used immediately, but is stored as potential energy in an air receiver tank in the form of compressed air.

The *safety-relief valve,* used to protect a system from excessively high or dangerous pressures, is a normally closed valve. The poppet of the safety-relief valve is seated on the valve inlet. A spring holds the poppet firmly on its seat. Air cannot pass through the valve until the force of the spring biasing the poppet is overcome. Air pressure at the compressor outlet is sensed directly on the bottom of the poppet. When air pressure is at an undesirably high level, the force of the air on the poppet is greater than the spring force. When this happens, the spring will be compressed and the poppet will move off its seat, allowing air to exhaust through the valve ports.

In order to control or regulate the air pressure downstream of the receiver tank, a pressure regulator is utilized. A pressure regulator is a pressure-reducing valve, and consists of a valve body with inlet and outlet connections, and a moving member that controls the size of the opening between the inlet and outlet. It is a normally open valve, which means that air is normally allowed to flow freely through the unit. With the regulator connected after the receiver tank, air from the receiver flows freely through the valve to a point downstream of the outlet connection. When pressure in the outlet of the regulator increases, it is transmitted through a pilot passage to a piston or diaphragm area, which is opposed by a spring. The area over which the pilot-pressure signal acts is rather large, which makes the unit responsive to the outlet pressure fluctuations. When the controlled (regulated) pressure nears the preset force level, the piston or diaphragm moves upward, allowing the poppet or spool to move toward its seats, thereby controlling the flow (increasing the resistance). The poppet or spool blocks flow once it seats and does not allow pressure to continue building downstream. In this way, air at a controlled pressure is made available to an actuator.

Control of Direction

In order to change the direction of airflow to and from a cylinder, we use a directional-control valve. The moving part in a directional-control valve will connect and disconnect internal flow passages within the valve body. This action results in a control of airflow direction.

A two-way directional valve consists of two ports connected to each other with passages that are connected and disconnected. In one extreme spool position, the flow path through the valve is open. In the other extreme, the flow path is blocked. A two-way valve provides an on-off function. This function can be used in many systems as an interlock and to isolate and connect various system parts.

A three-way directional valve consists of three ports connected through passages within a valve body: port A, port B, and port C. If port A is connected to an actuator, port B to a source of pressure, and port C is open to exhaust, the valve will control the flow of air to (and exhaust from) port A. The function of this valve is to pressurize and exhaust one actuator port. When the spool of a three-way valve is in one extreme position, the pressure passage is connected with the actuator passage. When in the other extreme position, the spool connects the actuator passage with the exhaust passage. Three-way valves may be used singly to control single-acting cylinders or in pairs to control double-acting cylinders.

One type of movable member quite frequently used to accomplish the direction change function is a poppet. This type of valve is shown in Figure 3.1. The poppet configuration is available in two-way, three-way, and four-way arrangements. However, to achieve the four-way function, two movable poppets are needed.

The movable member of the valve can be placed in several positions. The member is moved to these positions by mechanical, electrical, pneumatic, or manual means. Directional valves whose spools are moved by muscle power are known as manually operated or manually actuated valves. Various types of manual actuators include levers, push buttons, and pedals.

A very common type of mechanical actuator is a plunger equipped with a roller at its top. The plunger is depressed by a cam, which is attached to an actuator. Manual actuators are used on directional valves whose operation must be sequenced and controlled at an operator's discretion. Mechanical actuation is used when the shifting of a directional valve must occur at the time an actuator reaches a specific position. Directional valves can also be shifted with air. In these valves, pilot pressure is applied to the spool ends or to separate pilot pistons.

One of the common ways of operating a directional valve is with a solenoid. A solenoid is an electrical device that consists basically of a plunger and a wire coil. The coil is wound on a bobbin, which is then installed in a magnetic frame. The plunger is free to move inside the coil. When electric current passes through the coil of wire, a magnetic field is generated. This magnetic field attracts the plunger and pulls it into the coil. As the plunger moves in, it can either cause a spool to move or seal off a surface, changing the flow condition. When the motion of the solenoid is directly coupled to the shifting mechanism of the valve, it is called a direct-operated solenoid valve.

Another type of valve actuation is the pilot actuator. A pilot-actuated valve uses air pressure to move the valve spool. This air pressure may come from a variety of sources. Pilot-actuated valves are used where actuation is required in remote locations. They are also useful to control low pressures and are a requirement in pressure-centered valves. Since actuation forces increase the increasing pressure to the actuating ports, high forces can be generated to shift the valve members. Pilot-actuated valves may be internally or externally piloted. A valve is considered to be externally piloted if the air pressure received for shifting comes from an external source. If the pilot pressure comes from a source within the valve, it is said to be internally piloted.

Internally piloted valves use part of the pneumatic energy delivered to the pressure port to position the movable member within the valve

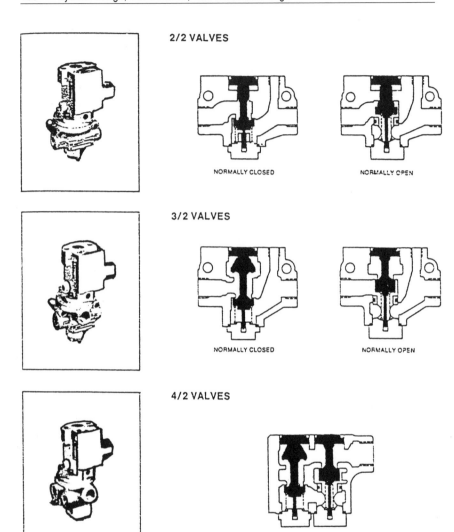

2/2 VALVES

NORMALLY CLOSED   NORMALLY OPEN

3/2 VALVES

NORMALLY CLOSED   NORMALLY OPEN

4/2 VALVES

Figure 3.1.   Poppet-type pneumatic direction control valves. (Courtesy of Ross Operating Valve Co., Detroit, Michigan.)

body. Such a valve has a definite minimum and, typically, a maximum pressure requirement. The application of this type of valve may impose circuit restrictions on the designer. This is due to the need to maintain the minimum specified pressure whenever the valve element requires shifting.

Conversely, externally piloted valves require a pressurized source that is derived externally to the valve. This means that the pressure can be selected in such a way that the valve will energize and deenergize in the same amount of time. If slow shifting times are required, a low pressure may be used. If rapid shift rates are necessary, we can use a high-pressure signal or the rapid exhausting of one of the pilot ports. However, when using high pressure to shift the main valve member, caution should be used. Extremely high velocities may cause high-impact forces on the valve member, leading to reduced life.

Control of Flow Rate

In a pneumatic system, actuator speed is determined by how quickly the actuator can be filled and exhausted of air. In other words, speed of a pneumatic actuator depends on the force available from the pressures acting on both sides of the piston, a result of the amount of air flowing into the inlet and out of the exhaust port.

A pressure regulator will influence actuator speed by portioning out to its leg of a circuit the pressure required to equal the load resistance at an actuator. This additional pressure is used to develop airflow. Even though this is the case, pressure regulators are not used to vary actuator speed. In a pneumatic system, actuator speed is affected by a restriction such as that obtainable with a needle valve or a needle valve with a bypass check, often called a flow-control valve. This valve will meter flow at a constant rate only if the system resistance, the total load (friction and load), and the pressure at the inlet of the cylinder and at its outlet do not vary throughout the entire stroke length. This valve does not control flow; it only affects flow. For example, when the load encounters an added resistance, actuator speed would decrease or the machine will stop. Once the resistance has been overcome, there would be a very rapid increase in speed. This could be a very dangerous condition for the operator or the machine.

As indicated, a needle valve in a pneumatic system affects the operation by causing a restriction. The typical needle valve consists of a valve body and an adjustable part. The adjustable part can be a tapered-nose rod, which is threaded into the valve body. The more the tapered-nose rod, is screwed toward its seat in the valve body, the greater the restriction to free flow.

By restricting exhaust airflow in this manner, a back pressure is generated within the actuator, thus reducing the forces available to create motion. This means that a larger portion of regulator pressure is used to overcome the resistances at the actuator, and less pressure energy is available to develop flow. With less air flowing into the actuator, actuator speed decreases. Again, by controlling the amount of restriction developed by a needle valve, the speed of an actuator can be controlled, but only if the total load is constant.

Maintenance and Troubleshooting

Pneumatic valves are rugged, reliable system components and are used extensively in various types of production equipment. This applies to metal-to-metal spool, resilient-seal spool, and poppet-type valve styles. However, best performance and long service life are influenced greatly by proper and regular maintenance. When malfunctions or failures occur, it is important for the maintenance personnel to be able to quickly determine the cause and source of the problem. Then the appropriate corrective action can be taken.

Maintenance

Following is a list of the maintenance and service requirements for proper valve operation.

1. **Supply clean air.** Experience has shown that foreign material lodging in the valve is a major cause of breakdowns. An air-line filter should be used that is capable of removing solid and liquid contaminants. Accumulated liquid should be drained from the filter frequently. If the filter is located where frequent routine maintenance is difficult, use a filter with an automatic drain.
2. **Supply lubricated air.** Most valves and the mechanisms they control require light lubrication. A good lubricator should put atomized oil into the air line in direct proportion to the rate of airflow. Either excessive or inadequate lubrication can cause the valve to malfunction. For most applications, an oil flow rate of 1 drop per minute is adequate. Another lubrication check can be made by holding a piece of clean white paper near the valve's exhaust port for three or four cycles. A properly lubricated valve will produce only a slight discoloration of the paper. Suitable lubricating oils must be compatible with the materials in the valve used for seals and poppets. Generally speaking, any light-bodied mineral- or petroleum-based oil with oxidation inhibitors, an aniline point between 82°C (180°F) and 104°C (220°F), and an SAE 10, or lighter, viscosity will prove suitable.

3. **Clean valve periodically.** The internal surfaces of a valve may gradually build up a deposit of varnish or dirt. This can lead to sluggish or erratic valve action, especially in spool valves. It is recommended that a schedule be established for the periodic cleaning of all valves. To clean valves, use a water-soluble detergent or a solvent such as kerosene. Do not scrape varnish surfaces. Also, avoid chlorinated solvents (trichloroethylene, for example) and abrasive materials. The former can damage seals and poppets, and abrasives can do permanent damage to metal parts.

4. **Clean electrical contacts.** In the external electrical circuits associated with the valve solenoids, keep all switches or relay contacts in good condition to avoid solenoid malfunctions.

5. **Replace worn components.** In many cases it is not necessary to remove the valve from its installation for servicing. Ordinarily, the only items that will need replacing after a long period of use are moving seals and possibly springs. These can be installed in the valve without removal from its mounting location. Before disassembling a valve or other pneumatic components, or removing it from its installation, shut off and exhaust the entire pneumatic circuit, and verify that any electrical supply circuit is not energized.

Troubleshooting

Troubleshooting is the process of observing a valve's trouble symptoms, such as a buzzing solenoid or sluggish action, and then relating these to their most likely causes. By careful analysis of the symptoms, the experienced troubleshooter can quickly determine the trouble, identify the cause, and take the appropriate correction action.

A pneumatic-valve troubleshooting chart (Table 3.1) is presented to assist in the process of identifying the cause and analysis of symptoms. The chart contains the most common trouble symptoms that valves experience and their probable causes.

Before disassembling a valve to investigate a system malfunction, check other possible causes of the malfunction. Because malfunctions in other components can affect valve action, the valve is sometimes blamed for a problem that, in fact, lies elsewhere. Leaky cylinder packings, poor electrical contacts, dirty filters, and air-line leaks or restrictions are just a few of the things to be considered when troubleshooting a pneumatic system. Consideration of these possibilities can sometimes

Table 3.1
Diagnosing Pneumatic-Valve Problems

Problem	Cause
Valve blows to exhaust when not actuated	1. Inlet poppet not sealing 2. Faulty seals 3. Faulty valve-to-base gasket 4. Cylinder leaks
Valve blows to exhaust when actuated	1. Faulty valve-to-base gasket 2. Faulty seals 3. Damaged spool 4. Cylinder leaks 5. Air-supply pressure too low 6. Water or oil contamination
Solenoid fails to actuate valve, but manual override does actuate valve	1. Loose pilot cover or faulty solenoid 2. Low voltage at solenoid
Solenoid fails to actuate valve and manual override also fails to actuate valve	1. Faulty seals 2. Varnish deposits in valve 3. Pilot pressure too low 4. Water or oil contamination
Airflow is normal only in actuated position	1. Broken return spring
Solenoid buzzes	1. Faulty solenoid 2. Low voltage at solenoid 3. Varnish in direct-operated spool valve
Solenoid burns out prematurely	1. Varnish in direct-operated spool valve 2. Incorrect voltage at solenoid
Pilot section blows to exhaust	1. Loose pilot cover 2. Pilot poppet not sealing
Poppet chatters	1. Air supply pressure low 2. Low pilot or signal pressure 3. Faulty silencer/muffler

Problem	Cause
Valve action is sluggish	1. Faulty seals on spool valve 2. Varnish in spool valve 3. Air-supply pressure low 4. Low pilot or signal pressure 5. Poor or no lubrication 6. Faulty silencer/muffler 7. Water or oil contamination
Sequence valve gives erratic timing	1. Faulty piston seal 2. Excessive lubrication 3. Fluctuating air pressure 4. Accumulated water 5. Faulty gasket
Flow-control valve does not respond to adjustment	1. Excessive lubrication 2. Incorrect installation or dirt in valve

save an unnecessary valve-disassembly job. The following paragraphs detail the corrective actions that can be taken for various causes of pneumatic valve failures.

Main inlet poppet not sealing. Foreign particles may be holding the poppet off its seat. Cycle the valve several times to see if the flow of air through the valve will flush out the particles. If not, it will be necessary to disassemble the valve (Figure 3.2).

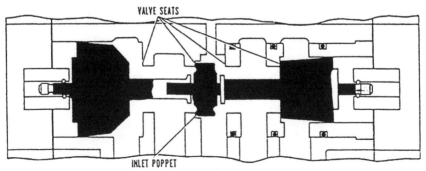

Figure 3.2.   Poppet valve with poppet not sealing. (Courtesy of Ross Operating Valve Co., Detroit, Michigan.)

To disassemble the valve, first disconnect or turn off the electrical circuit to the valve; shut off and exhaust the air supply; then disassemble the valve body assembly. The inlet poppet should be pulled off the valve stem and checked for dirt and damage. If damaged, the poppet must be replaced. If the poppet is swollen or has deteriorated, improper lubricants or solvents may be the cause.

Also, check the poppet seats for dirt and damage. If there is damage to a seat, the entire valve body assembly must be replaced (Figure 3.3). If there is no damage to the poppet or seats, clean the parts thoroughly, lubricate lightly, and reassemble.

Faulty seals. The materials of which seals are made can be attacked by substances such as chlorinated hydrocarbons (trichloroethylene, for example) and some lubricating oils. This can produce swelling or shrinking of the seals and result in erratic valve action or blowing to exhaust. Swollen seals may cause some in-line poppet valves to stick in a partially open position so that the valve blows to exhaust. Swollen seals on a spool valve can result in sluggish or erratic valve action, or even failure of the spool to move at all. Badly nicked or torn seals can produce blowing to exhaust in resilient seal-spool valves by allowing air to pass from one port area to another. Small leaks in piston-poppet seals can affect the timing accuracy of sequence adapters on in-line valves, or even render the valve inoperable.

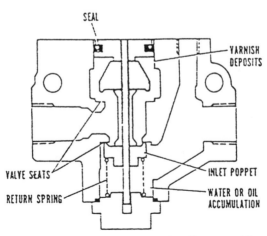

Figure 3.3. In-line-mounted poppet valve contamination. (Courtesy of Ross Operating Valve Co., Detroit, Michigan.)

Varnish deposits in valve. Varnish deposits can cause a valve to act sluggishly or even prevent movement of the valve element altogether, especially after a period of inactivity. A spool valve frozen in position by varnish can cause a direct-acting solenoid to buzz, and eventually leads to solenoid burnout (Figure 3.4).

Varnish results from the action of oxygen on the lubricating oil, and can be aggravated by excess heat. Varnish can also come from overheated compressor oil carried over into the air lines. Properly lubricated valves do not usually suffer from varnish deposits (Figure 3.5).

To remove varnish, use a water-soluble detergent or solvent such as kerosene. Do not scrape off the varnish. Also, avoid chlorinated solvents (trichloroethylene, for example) and abrasive materials. The former can damage seals and poppets, and abrasives can do permanent damage to metal parts. After cleaning, lightly lubricate moving valve parts and reassemble the valve.

Broken return spring. A broken return spring on a spool valve can cause the spool to remain in an actuated position, or to be only partially returned. In the latter case, several abnormal flow patterns may result, depending on the valve configuration. If a spool valve has a normal flow pattern only in an actuated position, a broken return spring is the most likely cause of the trouble. A broken return spring on an in-line poppet valve is less likely to prevent closing of the inlet poppet, but should be considered as a possible cause of the valve's blowing to exhaust when not actuated, especially in a low-pressure application.

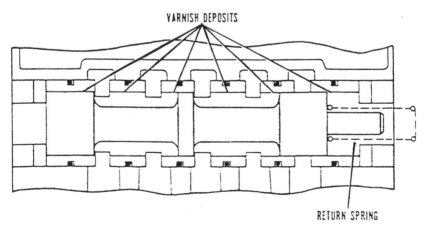

Figure 3.4.   Spool-sleeve valve varnish deposits. (Courtesy of Ross Operating Valve Co., Detroit, Michigan.)

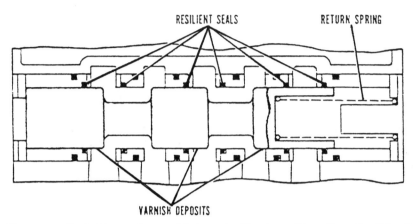

Figure 3.5.   Packed-bore valve varnish deposits. (Courtesy of Ross Operating Valve Co., Detroit, Michigan.)

Damaged spool. If a spool is badly scored or nicked, it can allow air to pass from one port area to another. This can result in unwanted pressurizing of an outlet port or blowing to exhaust. The problem can be further aggravated by the spool's cutting the resilient seals and increasing the leakage. A damaged spool cannot be repaired, but must be replaced.

Inadequate air supply. An inadequate air-supply volume causes an excessive pressure drop during valve actuation. Pilot air pressure may be great enough to begin movement of the valve element, but the pressure drop resulting from the filling of the outlet volume depletes the pilot air supply. This may result in chattering or oscillating of the main valve, or may simply keep the main valve partially actuated so that it blows to exhaust.

Check the pressure drop shown on the gauge at the pressure regulator. If the pressure falls more than 10% during actuation of the valve, the air supply may be inadequate. Inspect the system for undersize supply lines, sharp bends in the piping, restrictive fittings, a clogged filter element, or a defective pressure regulator. Remember, too, that the air volume supplied can be insufficient if more pneumatic devices are connected to a circuit than the compressor is designed to serve.

Fluctuating air pressure. If a valve with a timed sequence adapter suffers from erratic timing, the cause can be a fluctuating supply pressure. Consistent timing requires a consistent supply pressure. If the supply pressure varies considerably, install a pressure regulator set at the system's lowest expected pressure.

Inadequate pilot or signal pressure. Pilot or signal pressure below the minimum requirement can produce chattering, valve oscillation, or sluggish valve action. Check the valve specifications for minimum pilot or signal presure requirements.

Undersized or dirty silencer. An undersized silencer, or one that is partially plugged, restricts the exhaust flow. The resulting back pressure can cause erratic motion of valve elements. Remove the silencer to see if valve performance is improved. Clean the silencer to see if valve performance is improved. Verify that the silencer is of adequate size. Do not reinstall an undersized silencer. Install a larger silencer and check the valve performance again.

Lubrication. Some valves require lubrication to operate properly. Check the system lubricator to see that it is working as it should. Do not lubricate excessively. Excess oil can accumulate in low points of the system and restrict the flow of air. It can also form pools that will produce a dashpot effect and slow valve action. A visible oil fog exhausting from the valve is a sure sign of excessive lubrication. A properly lubricated valve will produce only a slight discoloration on a piece of white paper when held close to the exhaust port for three or four cycles.

Air cylinder leaks. Four-way valves sometimes blow to exhaust because of leaking packings in the air cylinder connected to the valve. Before looking for faults in the valve, check the cylinder for leaks. To check for cylinder leakage, the following steps can be taken (Figure 3.6):

1. Disconnect the air line to the end of the cylinder that is not under pressure. If air comes out of the open port, the cylinder packings are leaking and must be repaired. If there is no leakage, reconnect the line.
2. Reverse the position of the valve and disconnect the other air line to the cylinder. Again, check for air coming out of the cylinder port. If there is air coming out, the cylinder packings must be repaired.
3. If there is no leakage at the cylinder, reconnect the air line and proceed with troubleshooting of the valve itself.

Low applied voltage. Not enough magnetic force will be developed to allow the armature of the solenoid to seat. The unit will continuously

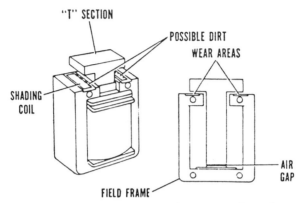

Figure 3.6. Checking for cylinder leakage. (Courtesy of Ross Operating Valve Co., Detroit, Michigan.)

draw high inrush current and burn out. Voltage should be checked at the coil with the solenoid energized. Possible causes of low voltage include high-resistance connections, too much load on the electrical circuit, and low voltage on the control transformer that powers the solenoid.

Valve spool or poppet stuck. The armature may be held unseated because the spool or poppet will not shift. The solenoid will draw high inrush current for too long a time and burn out.

A metal-to-metal spool type of valve may be varnished (stuck) in place, or dirt may prevent the spool from shifting. Clean, lubricate, and reassemble the valve.

A resilient-seal packed spool type of valve may not shift because swollen seals hold the spool in place, or because dirt prevents the spool from completely shifting. Clean the valve and repack the seals or replace the spool.

Solenoid corroded. Use valves with adequate protection against the moisture, coolants, and so forth, that may come in contact with them. Provide sealed electrical connections. Make sure that dirt and moisture covers are securely in place.

Solenoids energized simultaneously. On momentary contact, direct-actuated valves, check to make sure that both solenoids are never energized at the same time. This check is easily made by wiring a small indicator lamp temporarily across the coil of each solenoid. If both lamps are lit at the same time, the last solenoid to be energized will burn out. Correct the electrical circuit to prevent this.

Water or oil contamination. Accumulations of water or oil have an especially bad effect on devices with small orifices, such as timed sequence adaptors. Accumulations in such a device can change the effective size of the timing orifice or even block it completely. The device must be disassembled, cleaned, lightly lubricated, and reassembled. It may be necessary to install a filter in the supply line to prevent recurrence of the problem.

Accumulations of water or oil can also occur at low points in pilot supply lines. This can result in pressure fluctuations that produce erratic timing. The best cure is to reroute the pilot supply lines to eliminate low points. Water and oil can also accumulate at low points in a valve, and hinder movement of the valve element, perhaps completely preventing its motion. This is especially true of a valve operating in a sub-freezing environment where accumulated water can turn to ice. It is important in such applications to ensure that the supply air is dry, and that the air-line filter is drained frequently.

Internally piloted valve shifts improperly. An internally piloted valve may shift partially, then stall. Air blows steadily through the exhaust port. This is a sign that the pressure at the inlet port of the valve has fallen below the valve's minimum operating pressure. Increase the supply pressure or provide a local air accumulator to maintain pressure at the valve during periods of high flow.

If the valve is provided with an internal pilot exhaust, change it to exhaust externally if possible. Check for restrictions in the supply line. A gauge can be used to check the pressure available at the inlet port just before the valve fails to shift. Common causes of restricted supply lines are clogged filters, restrictive lubricators, and undersized hose or fittings.

On some valve designs, especially momentary contact types, full shifting may be obtained by restricting the exhaust port to allow the valve to maintain a pressure level above the minimum operating pressure. In extreme cases, provide a local accumulator for the pilot circuit or use pilot pressure from a remote source.

Valve occasionally malfunctions or circuit reaction slow. This type of problem can be caused by icing. Rapidly exhausting air often cools entrained moisture below its freezing point, causing ice particles to restrict exhaust flow. Mufflers can also freeze. Check the exhaust port of the valve and/or muffler for signs of icing. Dry the incoming air or provide a means for heating areas that tend to freeze.

Alternating-current solenoid failures. The failure of alternating-current (AC) solenoids can be due to a number of conditions. Figure 3.7 shows a typical pilot solenoid. Following is a list of the most common causes (Table 3.2).

High transient voltages. Solenoid burnout may be caused by high transient voltages that break down coil insulation, causing short circuits. High transient voltages are most common where solenoids are connected to lines operating above 120V, which also control motors and other inductive load equipment. Switching of such loads can create very high voltage peaks in the circuit. The remedy is to isolate solenoid circuits from main power circuits.

Temperature too high or too low. Solenoid failures can be expected when a valve is operated above its rated temperature. Insulation may fail because it is not suitable for high-temperature use. Specify solenoids designed for the ambient temperature, place the valve in a cooler location, or consider the use of a pilot-actuated valve at the hot location, controlled by a remote pilot valve.

Solenoids also can fail when valves are operated at lower-than-rated temperatures. Metal parts may shrink and lubricant viscosities may increase to a point where solenoid motion is retarded or stopped. This can be avoided by moving the valve to a warmer place or by providing a heated enclosure for the valve. If trouble persists with closely fitted valves, change to a type that can work at low temperatures.

Direct-current solenoid failures. Direct-current (DC) solenoids generally do not burn out due to the valve spool or poppet sticking, low voltage, or high voltage. This is because the DC solenoid does not have a high inrush current at the outset of the armature travel. The maximum current to the coil is the holding current, which is the same whether the solenoid armature is seated or not. With an AC solenoid, current to the coil at the outset of armature travel is three to four times greater than when it is seated (has completed its stroke). This absence of high inrush current in DC solenoids prevents coil burnout when the valve is stuck or the armature fails to move for other reasons.

Valve failures to shift can occur with DC as well as AC solenoids. Corrosion and temperature can also affect DC solenoid operation. The general guides for AC solenoids can be applied in these cases. High-transient-voltage problems can be handled in the same manner as for AC solenoids.

If the solenoid noise level on an AC solenoid is very high and occurs each time the solenoid is energized, check to see that the armature is

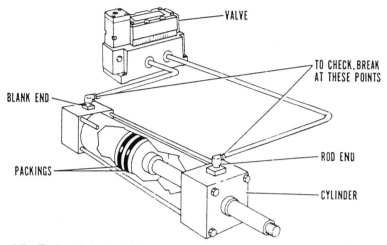

Figure 3.7. Typical pilot solenoid. (Courtesy of Ross Operating Valve Co., Detroit, Michigan.)

Table 3.2
Diagnosing Solenoid Burnout

Observable Damage	Possible Causes	How it Happened	Remedy
Coil insulation is burned; plunger is in open position; nylon coil bobbin is melted under the plunger	1. Low line voltage	Insufficient force to close plunger; high inrush current continues and generates excessive heat[a]	Replace coil only[b]; correct low voltage
	2. Ambient temperature too high	Plunger eventually will not close because undissipated heat has reduced current flow (closing force) while electrical resistance increases and generates more heat	Replace coil only[b]; install a continuous-duty model
	3. Cycling too fast		

(continued)

(*continued from previous page*)

Observable Damage	Possible Causes	How it Happened	Remedy
	4. Load too big or valve spool is blocked	Plunger blocked in open position permits electrical resistance to increase and generate excessive heat[a]	Replace coil only;[b] correct high-load condition
Coil insulation is burned; plunger is in closed position	1. Voltage too high 2. Solenoid too large for light load	Extra force of excessive holding current holds plunger in close position while electrical resistance increases, generates excessive heat, and burns out coil	Replace coil only[b]; correct high-voltage condition
Frayed and burned lead wires	1. External mechanical short	Water-based coolant, metal chips, etc., have created contact between wires	Replace coil only[b]; shield from coolant
Small pinhole burn in coil wrap	1. Transient short to ground	High-voltage surge causes spark to jump between coil winding and solenoid C-stack (or other nearby ground)	Replace coil only[b]
Spongy insulation on coils and lead wires	1. Internal mechanical short	Fire-resistant fluids (phosphate esters) dissolve coil insulation, coil varnish, paint, etc.; cause short between coil turns	Install solenoid with proper insulation

Observable Damage	Possible Causes	How it Happened	Remedy
Deep scoring at all seating surfaces	1. Over-voltage or reduced load (or wrong size)	Excessive closing force causes T-bar to wear through copper shading coils at top of C stack; plunger also hammers base, resulting in destruction	Replace the entire solenoid, not just the coil

[a] *Excessive heat burns insulation off coil wires, permitting electrical short and melting nylon bobbin.*

[b] *If solenoid is the encapsulated type, the entire solenoid must be replaced.*

NOTE: In the case of double-solenoid valves, the cause of solenoid burnout may be that both solenoids were actuated simultaneously. Usually, one solenoid will burn out from inability to close properly.

* Courtesy of Parker Hannifin Corp., Cleveland, Ohio.

seating. Most direct-solenoid-actuated valves are provided with a manual override. If the solenoid noise decreases when the override is operated, incomplete solenoid motion is indicated. Check to determine if rubbing parts can move properly and that the correct voltage is available at the coil. Extremely loud AC hum can be caused by a broken part within the solenoid.

In marginal noise problems, consider mounting the valve so that the armature works up and down rather than horizontally. Maximum quieting can be obtained by using DC solenoids. This is practical even on large valves of the solenoid-controlled, pilot-actuated type. A flowchart summarizing the solenoid-valve troubleshooting procedure is presented in Figure. 3.8.

Air Motors

Although not as widely known and applied as hydraulic motors, rotary air motors, using a compressed gas (in place of a pressurized liquid) as their power source, provide the plant engineer with rugged, inexpensive drives requiring little more than shop air for their operation.

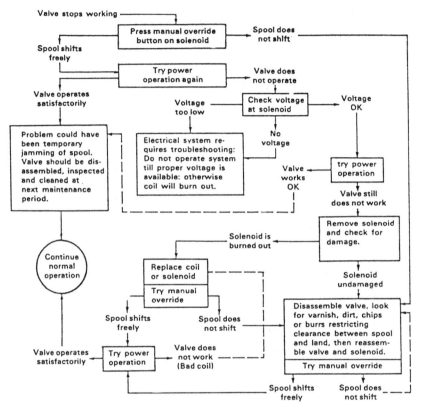

Figure 3.8.   Troubleshooting guide for single-solenoid direction-control valve. (Courtesy of Parker Hannifin Corp., Cleveland, Ohio.)

The term "air motor" can refer to two quite different types of pneumatic mechanisms. Reciprocating air motors that provide a hammering motion find only limited application in industrial plants, largely in air-driven handtools. The more versatile rotary motors, such as electric or hydraulic motors, produce rotation of a shaft. They have special advantages whenever speeds above 3,600 rpm or below 850 rpm are required, since they are capable of speeds far in excess of those normally expected of motors of other types, and (like hydraulic motors) reach their highest torque when nearly stalled (Figure 3.9).

Many designs are fully reversible, a capability that is particularly useful when the motor is to be employed to position a load (Figure 3.10), and

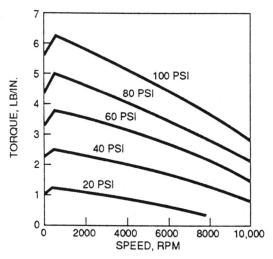

Figure 3.9. Air-motor torque.

Figure 3.10. Vane motor-powered hoist. (Courtesy of Ingersoll-Rand Corporation.)

most can be operated vertically or horizontally. Because there is no shock hazard associated with these motors, they find application in construction projects, where powered equipment must be used under wet conditions. Also, because there are no commutator-ring sparks, they are often used to power mixers and other equipment that must operate in explosive or potentially explosive atmospheres.

Although the compressed air supplied to the motor must be clean, air motors are tolerant of dirty exterior conditions. In contrast to electric motors, they can be stalled indefinitely without heat buildup and without damage. During the stall period, they continue to torque their shafts, so that when the load is reduced, or the air presure increased, the motor goes into operation again without operator attention.

Continuous duty, however, is not normally recommended. Because of compressor and pneumatic line losses, as well as inefficiencies in the air motors themselves, they are more costly to operate than many other types of motors. Also, most air motors employ sliding parts, which wear comparatively rapidly even when well lubricated.

Noise is likely to be a problem with air motors. If the motor is small, it may be practical to muffle the exhaust at the outlet. To avoid reduction in efficiency associated with muffling, larger motors may have their exhausts piped to an area remote from personnel.

Power is derived principally from the kinetic energy of the compressed air, not from its potential energy, though some expansion in the motor may take place. The operation of a pneumatic cylinder furnishes a good analogy: When the rod has been extended and the work of the cylinder has been accomplished, the pressure in the cylinder is still equal to the pressure of the air line to which it is connected. The vanes or pistons of an air motor operate similarly.

Variable-displacement construction, frequently found in hydraulic motors, is not available in air motors, since use of a compressible medium makes it unnecessary. When high speed under light loads with low power consumption is desired, it can be achieved by throttling the air supply.

Types of Air Motors

Vane motors have a rotor mounted eccentrically in a cylinder, with longitudinal vanes fitted into radial slots running the length of the rotor. Torque is developed from the pressure imbalance acting on the vanes, and the user can increase torque at any speed by increasing air pressure at the inlet. But there are limitations—cost and available pressure (Figure 3.11).

There may be from three to ten vanes per motor: the more vanes, the less blowby (internal leakage). As the number of vanes is increased,

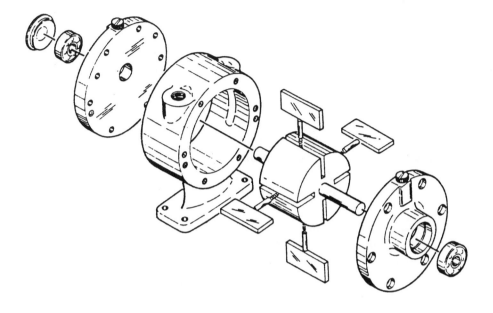

Figure 3.11.   Vane air motor.

starting torque and reliability are also improved. So is motor cost—in the manufacturer's favor.

If a motor has only three vanes, the sticking of one vane in the retracted position can prevent the motor from starting under load. Using four or more vanes—plus porting line air to the bases of the vanes—overcomes this problem. There are special types of air motors that require specific numbers of vanes, but generally the number of vanes is not critical unless uniformity of torque at low speed is important.

Vane motors are usually reversible. Compressed air entering the motor's cylinder exerts equal pressure on the power-side and returning vanes; but because the returning vanes are retracted, the resulting forces are unbalanced, and the rotor revolves. Thus, the force at stall is equal to the product of the inlet pressure and the difference in area between the power-side and returning vanes, and the torque is proportional to the diameter of the cylinder.

Foot-, hub-, face-, and flange-mounting vane motors are available. Typically, rotor and cylinder are machined from fine-grained cast iron.

The vanes are usually of nonmetallic material to provide the lowest practical coefficient of friction with the cylinder wall and to reduce mass.

In air motors, starting torque is as important as maximum torque, which is reached just above zero speed and decreases linearly as the motor approaches free speed. The starting torque is generally 75 to 85% of maximum torque because of internal static friction. Starting torque is the limiting factor of a motor, since the motor must overcome the applied load if it is to be useful.

In vane motors, effective torque output is proportional to the exposed vane area, pressure imbalance across the blade, and the moment arm through which the pressure is working. Torque at any given speed can be increased by:

1. Increasing the exposed vane area.
2. Increasing inlet air pressure, thereby increasing pressure imbalance across the blade.
3. Increasing housing inside diameter (ID) or length and/or rotor outside diameter (OD) or length.

The amount that any one of these can be increased is limited. Increasing exposed vane area increases porting problems and shortens blade life.

Vane motors operate at motor speeds from 3,000 to 25,000 rpm, and frequently come equipped with governors to prevent overspeed when loss of load occurs. If a nongoverned multivane motor operates in a no-load condition, its high running speed can char the vanes as they rub the cylinder wall, and damage other motor parts.

Vanes will wear even at moderate speeds, yet with proper maintenance will provide several hundred hours of trouble-free operation. Lubrication is by airborne oil mist. Multivane air motors are inherently higher speed units and deliver more power per pound of weight than piston motors, but require more servicing.

Lubrication is frequently a problem when short duty cycles are followed by long periods of inactivity. Under these conditions, motor life will be lengthened if a lubricator is placed as close as possible to the motor. This type of motor finds application as a drive for mixers, pumps, fans, tools, conveyors, hose reels, shop hoists, feeders, rewind stands, and similar devices.

Piston motors have more positive starting, better speed control at low speeds, and slightly better stopping characteristics than vane motors. In addition, they have slightly lower air consumption at the lower speeds.

Output torque is developed by pressure acting on pistons; there are 4, 5, or 6 pistons per motor. Most motors have 5 or 6 pistons, but there is some disagreement as to which number is preferable since both designs give excellent performance.

Power developed by piston motors increases with inlet pressure, number of pistons, piston area, stroke, and speed. Normally, piston motors are not equipped with governors to prevent overspeeding.

The primary factor limiting speed in a piston motor is the inertia of the moving parts. This inertia has a greater effect in radial-piston motors than in axial types.

Axial-piston air motors, Figure 3.12, are both more complex and more costly than vane motors. They are also smoother running, and produce their maximum power at much lower speeds. The cylinders and cylinder block do not revolve as they do in a hydraulic swashplate motor, which the axial-piston air motor otherwise resembles. Instead, the same motion that drives the cam plate and, through gearing, the output shaft, causes the inner-port control valve to rotate, alternately pressurizing and exhausting the cylinders. Axial-piston motors are usually smaller and lighter than electric gear motors of the same horsepower, will tolerate higher ambient temperatures (typically an average of 200°F), and can be lugged down to stall repeatedly without damage.

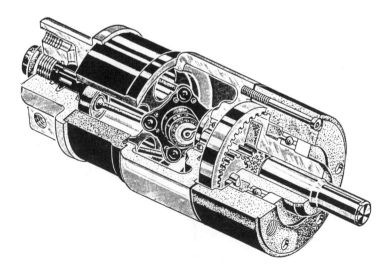

Figure 3.12. Axial-piston air motor.

Radial piston motors, Figure 3.13, have the highest starting torque of any type of air motor, and are thus particularly suited to applications involving heavy starting loads. They are reversible and, although their maximum speeds are usually lower than those of other types of air motors (2,000 to 3,600 rpm), geared types are available. The pistons surround a crankshaft, which is coupled to the valve block; compressed air from the inlet port is directed by this rotating valve to the proper piston—in most designs two pistons are delivering power strokes at any given time. When each piston has reached the end of its stroke, the valve is in the correct position to exhaust the air.

The radial-piston motor is the most common of the two types of piston motors. The axial type is more compact than the radial, but because it is grease-lubricated, it requires more servicing. Axial-piston motors are not generally offered in sizes over 3½ hp.

Radial-piston motors have better starting characteristics than axial-piston motors. Basically low-speed devices, they operate with free speed up to 3,800 rpm. They can lug heavy loads at all speeds and are particularly adaptable to applications requiring high starting torque and slow speed. They have built-in splash oilers, and normally must be installed with the crankshaft in a horizontal position; vertical mounting causes lubrication problems that are expensive to correct.

Figure 3.13. Radial-piston air motor.

Other types of air motors include the V-motor, a type of radial-piston motor in which pairs of cylinders, arranged in a V-configuration, drive a crankshaft. Similar in concept to automotive V-block engines, they are grease-lubricated, can be operated in any position, and are narrower than radial piston designs of similar hp. Like radial-piston motors, they have good lugging characteristics, but they are more expensive, and have been largely phased out in the interest of economy. One continuing application for V-motors is as feed motors for large rock drills.

Turbines, with their very low starting torques, are used for extremely high-speed applications, such as jet aircraft engine starters and high-speed, light-load grinding. The use of turbines larger than those required for small grinders is limited, and each application requires a custom design. Generally, turbines are not available from the manufacturers of air motors, but from engine and accessory manufacturers of air motors in the aircraft and missile fields.

Diaphragm motors use a reciprocating diaphragm and a ratchet mechanism to obtain rotary motion. These motors offer high torque at low speeds and are used in applications where torque requirements vary widely.

Air-Motor Characteristics

Power characteristics of air motors are similar to those of series-wound DC motors. At constant inlet air pressure, the brake hp is zero at zero speed, increasing with increasing speed until it peaks at approximately 50% of free speed. It then decreases to zero again at free speed (maximum speed under no-load condition).

Air motors operate principally by the force of air pressure working on the vanes or pistons. While they also obtain a portion of their power from the expansion of compressed gas, expansion is, by no means, the main source of energy because it is not practical. When air or gas expands, it cools. Excessive expansion would cause condensation to freeze in the exhaust ports, choking the motor. The maximum amount of expansion that can be safely used is about 20%.

Torque reaches its highest value slightly beyond zero speed and falls off rapidly, almost in a straight line, until it reaches zero at free speed. The maximum torque may vary from 0.6 to 1,175 ft-lb, depending on the type and size of the motor. Torque when stalled is approximately twice the torque at rated horsepower. Stall torque can be determined readily from the horsepowers and speeds listed in manufacturers' literature.

Starting torque is the maximum torque the motor can produce when starting under load; its value is approximately 75% of stall torque. Because static friction at the vane tips exceeds dynamic friction, it takes more force to start a vane motor than to keep it running. If the load on the motor exceeds its starting torque, the motor will not start.

Rated horsepower normally refers to the maximum power produced at 90 psi. Available air motors range from 0.12 to 25 hp, although not all types can be obtained in all sizes. Vane motors are built in the entire range of ratings; axial-piston motors from ½ to 3½ hp; radial piston-units from 1½ to 25 hp.

Standard air motors are built for operation on 30 to 150 psi air at the motor intake, but most are operated in a narrower range of 60 to 100 psi. If the load requirements fall outside this bracket, most manufacturers recommend substituting a different model.

Comparisons can be made among motors rated at different pressures by allowing 14% change in hp for each 10 psi pressure change. Thus, a motor rated at 8 hp at 90 psi would be equivalent to one rated at 6.9 hp at 80 psi. By the same rule, a 10 psi drop in air inlet pressure will decrease the motor's efficiency by 14%. This directly affects cost and productivity.

Control of air pressure supplied to the motor is the simplest and most efficient method of changing the operating characteristics of an air motor. Conversely, failure to maintain the desired air pressure at the motor inlet is the most certain way to nullify the design characteristics of the motor.

If a motor is rated at 90 psi, it is not enough to determine that there is 90 psi at the compressor. There must be 90 psi at the motor for it to perform at its rated torque and hp.

Torque and hp for any air pressure can be plotted on curves similar to Figure 3.14. Normally, manufacturers show only three or four different pressures on a chart. Torque and hp can be roughly interpolated for other air-inlet pressures.

Speeds at rated hp range from 38 to 6,000 rpm. These power and speed ranges do not have a direct relationship; in other words, the lowest hp does not indicate the highest speed or vice versa.

Free speed is the maximum speed under a no-load condition; for a governed motor, the term free speed actually means free governed speed, i.e., the maximum speed at which the motor will run with the governor operating. Free speed normally varies from 60 to 13,000 rpm for vane motors and 190 to 4,600 rpm for piston motors, although

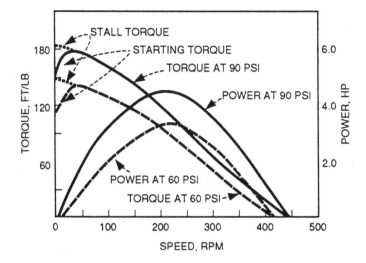

Figure 3.14.   Air-motor torque, hp, and speed relationship.

speeds can range from zero to 25,000 rpm to match practically every requirement for a compact source of smooth power.

Design speed, that point at which rated or maximum power is reached, is approximately one-half of free speed for a nongoverned motor and about 80% of free speed (free governed speed) for a governed motor. For the most efficient operation, a motor should run at design speed.

Increasing the load will decrease the speed of an air motor, slowing it down until its torque equals the load requirements. By opening the throttle to increase the air pressure, the motor then can be brought up to rated speed.

Selection factors, used to decide between vane and piston motors, vary in importance. First consideration should be given to the desired operating speed, not free or rated speed. When the operating speed will be less than 25% of free speed, use a piston type. Good speed control at slow speeds is characteristic of piston motors, and they have slightly lower air consumption than vane motors at slower speeds. Service, maintenance, and actual hours of operation are more important than the duty cycle of an air motor. The length of time a load is applied to an air motor incorporated in stationary equipment is not a major concern.

If the motor will be used under varying loads, the major factor affecting selection is whether or not the motor has sufficient speed and torque to satisfy the operating conditions. Two motors may produce the same maximum hp, but have substantially different speeds at the same loads.

Most air motors, of necessity, are geared motors. Gearing is used to reduce the high speeds of vane motors to more usable speeds. Since maximum power normally occurs at approximately one-half of free speed, reducing free speed also reduces design speed. But gearing also reduces efficiency, so the less gearing the more efficient the motor.

Operating characteristics of air motors are accurate, rapid, and reversible speed and power control. They start and stop instantaneously and will run at an infinite number of speeds from a slow crawl to full speed. Controls necessary to obtain this wide range of speeds are simple, inexpensive, trouble-free and compact.

Reversing operation. One of the major advantages of an air motor is its ability to come up to full speed almost instantly. Vane motors with no connected inertia come up to speed in one-half revolution. Piston motors are even faster and reach full speed in milliseconds.

Valves control the direction of rotation and the speed of air motors. They may be hand, foot, pilot air, solenoid, or pendant operated. Whatever the type, the valve should have full-flow air passages in order to take advantage of the full power of the motor. A rotary four-way valve is the standard control for both throttling and motor reversing.

If the motor is nonreversing, it can be controlled with a globe valve. Either valve can be used where infrequent speed adjustment or constant speed setting is desired, or where remote actuation is necessary. An air motor may be regulated from minimum to maximum speed without harming the motor, and reversed rapidly without damage.

Governors are available for both piston and vane motors, although normally they are not furnished with piston motors unless specified. Nongoverned motors can be damaged by operating them in a free speed (unloaded) condition. Frequent shock loading will also damage various motor parts. If the motor is applied in such a way that it cannot be unloaded, a governor is unnecessary.

Governed air motors produce maximum power at approximately 80% of free governed speed.

Air consumption varies with speed and motor size. Two motors operating at 90 psi may consume anywhere from 21 cfm to 45 cfm/rated hp, with the larger models having the lower consumption-to-hp ratios because of better porting.

When estimating air consumption per minute, allow 34 to 40 cfm/ hp at maximum hp for motors below 1½ hp, and 25 to 30 cfm/hp for larger motors. To estimate air consumption for a particular speed, determine the consumption at rated speed, at ¼ free speed, and at free speed. Manufacturers usually supply the air-consumption rate for rated speed or free speed or both. Once air consumption at the three speeds has been determined, fit a smooth curve through them. To ascertain total air consumption for a particular operation, multiply consumption per minute at the operating speed by cycle time (working time).

Air-Motor Selection

At times, cost or motor size will dictate the type of air motor chosen. For example, vane motors are the only choice for portable power tools; the bulk and weight of a piston motor make it impractical for use in handheld tools (Figure 3.15).

For hoists and winches, piston motors should be selected because of their good start and stop characteristics and their ability to handle heavy loads at low speeds. Another important advantage is that trapped air in the compression and exhaust cycle acts as a brake on the load, providing an additional safety factor.

Because the performance range of vane and piston air motors overlaps nearly 100% (Table 3.3), in many instances factors other than performance will decide the choice. One of the most important is required gearing. When either type of motor will perform the desired task and the differences cited are not of prime importance, the user normally should select the motor that requires the least gearing, since gearing reduces efficiency.

Maintaining Air Motors

Many air motors in use today are operating 30% or more below their rated efficiency, because of improper care and handling. Therefore, more attention to specific factors governing efficient operation of the air system is required. Every motor in the plant should be regularly inspected, cleaned, and maintained. This inspection program should include provision for determining whether motors have become uneconomical to maintain and, if this happens, should be replaced by new, efficient units.

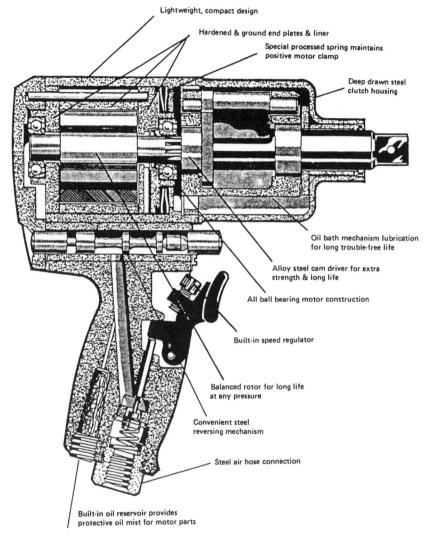

Lightweight, compact design

Hardened & ground end plates & liner

Special processed spring maintains positive motor clamp

Deep drawn steel clutch housing

Oil bath mechanism lubrication for long trouble-free life

Alloy steel cam driver for extra strength & long life

All ball bearing motor construction

Built-in speed regulator

Balanced rotor for long life at any pressure

Convenient steel reversing mechanism

Steel air hose connection

Built-in oil reservoir provides protective oil mist for motor parts

Figure 3.15. Vane air-motor-driven impact wrench.

There are six basic conditions that can decrease air-motor efficiency and cause a sharp rise in operating costs:

1. Excess moisture in the motors
2. Foreign particles abrading internal parts
3. Missing strainers/filters

Table 3.3
Typical Air-Motor Performance Ranges

Maximum Horsepower	At Max. HP rpm	Maximum Torque		Nonreversible or Reversible
		ft lb	mkg	
Multivane Motors				
Flange-, Foot-, or Face-Mounted				
.25	1,800	1.1	.15	R
.50	1,800	2.3	.31	R
1.2	2,000	4	.55	R
2.1	2,000	8	1.1	R
5.0	1,800	20	2.8	R
7.5	1,800	26	3.6	R
Flange-Mounted				
.12–.15	150–1,500	1.0–7.3	.14–1.01	R
.18–.20	150–1,600	1.3–13.0	.18–1.80	R
.22–.25	275–900	2.4–8.5	.33–1.18	NR
.34	225–1,350	2.6–16.0	.36–2.21	NR
.45	38–6,000	.6–97	.08–13.42	R
.70	57–1,080	6.3–120	.87–16.60	NR
1.1	150–430	20–60	2.8–8.3	R
1.4	175–5,000	23–70	3.2–9.7	NR
1.4	98–390	32–115	4.4–15.9	R
1.5–1.6	121–520	33–118	4.6–16.3	NR
3.0–3.2	125–485	65–215	9.0–29.7	R
3.5–3.7	145–560	70–240	9.7–33.2	NR
4.5	4,100	10	1.4	NR
5.6	47–95	545–1,110	75.4–152.1	R
5.7	42–85	580–1,175	80.2–162.5	NR
7.4	1,800	39	5.4	R
8.6	1,800	45	6.2	NR
20	780–2,200	77–200	10.7–27.7	R
24	1,000–2,580	88–225	12.2–31.1	NR
Base-Mounted				
1.5–1.6	170–1,150	12.5–79	1.7–10.9	R
1.9	280–1,250	13–58	1.8–8.0	NR
2.5–2.7	55–520	55–470	7.6–65.0	R

(continued)

(continued from previous page)

Maximum Horsepower	At Max. HP rpm	Maximum Torque		Nonreversible or Reversible
		ft lbs	mkg	
3.3–3.6	65–225	145–485	20.1–67.1	NR
3.4	100–285	110–300	15.2–41.5	R
5.6	47–235	215–1,100	29.7–152.1	R
5.7–6.0	42–215	230–1,175	31.8–162.5	NR

Piston Motors

Base- or Flange-Mounted				
1.8	1,800	7.8	1.1	R
2.1	2,500	7.5	1.0	NR
2.6	1,250	21	2.9	R
2.9	1,400	19	2.6	NR
3.1–3.4	850–1,075	28–34	3.9–4.7	R
3.8–4.0	1,250–1,400	26–32	3.6–4.4	NR
4.7–7.2	92–925	62–600	8.6–83.0	R
5.4–9.8	130–1,525	55–530	7.6–73.3	NR
9.5–11	650–700	125–156	17.3–21.6	R
12–14	1,075–1,100	115–145	15.9–20.1	NR
15–21	625–680	240–320	33.2–44.2	R
19–25	1,000	185–260	25.6–36.0	NR

4. Improper lubrication
5. Abuse and misapplication
6. Failure to replace worn parts

An air-motor troubleshooting checklist is presented on Figure 3.16 to assist with determining the cause for a variety of problems.

Some of these basic problems are attributable to air-system design but may be minimized by planned, periodic, preventive maintenance.

Clean, moisture-free, and well-lubricated air at proper pressure is the first and most essential requirement for efficient operation of a pneumatic motor. Dirt and scale in the air system cause wear of fitted parts, and water washes away lubricant, causes corrosion and, because it freezes when the air expands, can cause loss of power.

Because most air motors are designed to operate at 90 psi (at motor inlet) for maximum efficiency, adequate air supply is a major factor

AIR MOTOR TROUBLESHOOTING CHECKLIST	Air Leakage	Air Strainers Clogged	Air Pressure Too Low	Dirty Air	Water In Air	Incorrect Lubrication	Insufficient Lubrication	Hose Too Small	Long Vanes	Worn Vanes	Rotor Rubbing	Worn Bearing Plates	Worn Valve Seat	Throttle Pin Sticking
Motor Won't Run		●	●				●		●		●			
Lack of Power	●	●	●		●			●		●	●	●		
Speed Too Low		●	●					●			●			
High Air Consumption	●									●		●		
Excessive Vane Wear				●		●	●							
Excessive Bearing Wear				●		●	●							
Rusting of Parts					●	●	●							
Delamination of Vanes				●	●	●								
Vanes Chipping				●		●	●							
Motor Continues to Run, Throttle Off													●	●

Figure 3.16.   Air-motor troubleshooting checklist.

in efficient air-motor operation. A drop of 10 psi at the inlet reduces the motor's output by 14%; a drop of 30 psi could affect output by as much as 45%. Leaking air lines, small capacity hoses, and inadequate compressors can defeat the best motor design.

Typically, each component of an air motor—vane, piston, rotor, cylinder, etc.—is individually replaceable. The motor should be taken apart periodically and cleaned with solvent. Dirt and gums should be removed from all operating parts, including the motor, governor, and clutch, and the parts checked for wear.

Strainers are furnished with most quality pneumatic tools and motors. They are located at the inlet ports, to prevent particles of dirt and rust from entering the motor and causing severe abrasive wear. Strainers should be checked and cleaned regularly. After cleaning (this cannot be stressed too strongly), they must be returned to the motor.

Pneumatic motors must be lubricated to prevent needless wear, although some of them, particularly vane motors, are equipped with oil reservoirs. Each time the motor stops, air pressure in the reservoir is greater than that in the rest of the motor, and as the air exhausts through the outlet ports, a small amount of oil is carried with it.

By holding a piece of plain white paper in front of the exhausting air stream, the operator can determine whether or not the feed rate from the oil reservoir is properly adjusted. If fine spray appears on the white paper, then the motor is in proper adjustment.

Motors not equipped with oil reservoirs must be lubricated by in-line oilers or lubricators. The lubricator should be located at the point where the air hose connects to the supply lines, or as close to this point as possible. One oiler or air-line lubricator should be used for each pneumatic motor on the line, when continuous duty is expected, or the motors are permanently connected.

A tight fit is necessary when installing couplings. The nipple and inside of the hose should be lubricated with a soapy solution if fitting is too difficult. Never use oil or grease on couplings, or ream the hose to facilitate fitting; the end must be cut squarely. After the coupling has been inspected for burrs and serrations it can be placed in a vise and the hose forced over the sleeve as straight as possible. Hammers of all types should be avoided.

Piston motors.   Lubrication is the most critical phase of piston-motor maintenance. The built-in chamber must contain oil. Because condensation of moisture occurs, this splash chamber must be drained and refilled on a regular schedule determined by the application.

Some piston motors are equipped with an oil-reservoir vent plug. When this plug is lost, a solid plug is often put in its place. This plug causes pressure buildup in the oil reservoir, and oil is forced out, quickly depleting the supply. Leaving the vent plug out altogether is equally bad practice; dirt and grime fall into the oil reservoir and accumulate. When the vent plug is removed, it should be retained, and replaced immediately after servicing, so that it won't be lost. On most quality motors, this plug is attached with a chain.

Vane motors.   Proper performance of these motors requires an unrestricted flow of air at the recommended pressure, proper lubrication and filtration, and the use of recommended replacement parts.

Vanes are the heart of the motor. The leading edge of each vane must be in excellent condition if the vane is to seal properly. Excessive wear on the leading edge is usually an indication of insufficient oil in the air system.

Chipped, cracked, loose, or rough-edged vanes should be replaced to maintain the efficiency of the motor and the machine or tool it drives. When the width of the vane has worn excessively, the vane should be removed and a new one inserted, for narrow or thin vanes can break and wedge between the rotor and cylinder, causing extensive damage.

The Sources, Effects, and Control of Contamination in a Hydraulic System

Nearly all hydraulic-component manufacturers and users agree that *dirt* is responsible for a majority of malfunctions, unsatisfactory component performance, and degradation. This dirt comes from many sources and in various forms, such as chips, dust, sand, moisture, pipe sealant, weld splatter, paints, cleaning solutions, and others. The aggregate of all these forms is termed *contamination,* rather than dirt.

Sources of Contamination

The sources of contamination are many (Figure 4.1). Contamination can be

- Introduced with new oil
- Built-in at the time of hydraulic-system construction
- Introduced with the air from the environment
- Generated by wear within the hydraulic components
- Introduced through leaking or faulty seals
- Introduced by shop maintenance activities

The removal and control of contamination necessitates the use of a filter. The selection of the correct filter and its proper location in the system requires as much care and expertise as the selection of other critical components such as pumps, valves, and actuators.

Contamination is introduced into hydraulic systems at the time the components are made and during the manufacture of the system. Some of the contamination found in oil samples taken from a system after a short run-in period of a new machine includes:

- Metal chips from tubing burrs, pipe threads, component manufacture particles, and tank fabrication

171

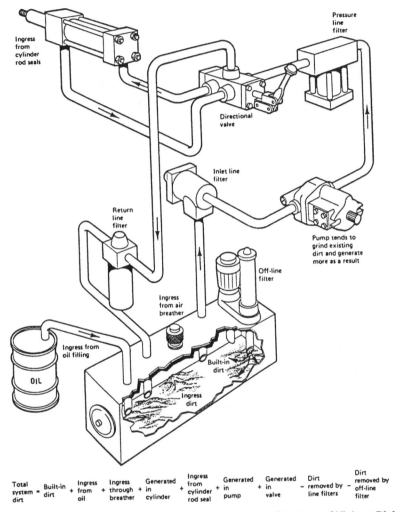

Figure 4.1. The sources and control of contamination. (Courtesy of Vickers Division, Trinova Corporation.)

- Pipe dope
- Teflon tape shreds
- Lint from wiping cloths
- Welding scale and beads
- Pipe scale
- Factory dust

■ Rubber particles
■ Debris from hose fabrication

Although oil is refined and blended under relatively clean conditions, it is usually stored in drums or in a bulk tank at the user's factory. At this point, it is no longer clean because the filling lines contribute metal and rubber particles, and the drum always adds flakes of metal or scale. Storage tanks can be a real problem because water condenses in them, which causes rusting, and contamination from the atmosphere finds its way in unless satisfactory air-breather filters are used.

If the oil is being stored under reasonable conditions, the principal contaminants on delivery to a machine will be metal, silica, and fibers. Using a portable transfer unit or some other filtration arrangement, it is possible to remove much of the contamination present in new oil before it enters the system and is ground down into finer particles by the hydraulic equipment.

Contamination that enters the oil from the environment surrounding the hydraulic system can do so by following several paths. These entry points include air breathers, access plates, and seals.

Air breathers. Very little information appears to be available on what these filters will actually achieve, and purely nominal ratings are ususally specified. The amount of air passing through the filter will depend on draw-off, which means that single-acting cylinders operating in dirty atmospheric conditions will result in a greater introduction of dirt.

Access plates. In some plants it cannot be assumed that access plates will always be replaced. In power unit design, good sealing is vital, and in bad environments such items as strainers should not be positioned inside the reservoir if access requires the refitting of removable plates. Other removable items will allow dirt to enter during maintenance, and good design practice should minimize this.

Seals. Wiper seals in cylinders cannot be 100% effective in removing very fine contaminants from the cylinder rod. If they were, they would remove the oil film from the piston rod. A completely dry rod would quickly wear out the seals. Where cylinders remain extended in a heavily contaminated atmosphere, considerable quantities of fine particles can get into the system unless protection such as a bellows is provided.

Dirt is continually introduced into operating hydraulic systems because of wear and degradation of the working components. The wearing action of working parts in components such as pumps, fluid motors, valves, and cylinders generates contamination. Rust scale from the reservoir caused by condensation above the oil level is also a source of dirt. Burrs on tubing and piping break loose during service; the flexing of components continuously releases particles that were not removed during the initial cleaning of the system.

Contamination is added to the system by shop maintenance activities. This is done in various ways including the following:

- Leaks are caught in a bucket, which may be dirty and open to the atmosphere, then returned to the tank.
- Dirty buckets are used to catch oil when components are repaired or lines broken then returned to the tank.
- Oil is stored in drums that are not sealed.
- Occasionally, supposedly clean oil, which is new or reprocessed, will contain lint particles from the refining filters.
- When equipment is being repaired
  a. Components pick up contamination from dirty workbenches.
  b. Dirty rags are used to clean components.
  c. Tubing and hose is left on the floor or exposed to dirt.
- Suction-line strainers are often left lying in the tank, or are perforated with a long screwdriver by maintenance people who "don't want to be bothered" cleaning strainers under oil or who are in a hurry to get the machine going.
- Strainers are taken off for cleaning. The procedure stirs up the dirt in the bottom of the tank. Pumps are started up because "it won't hurt to run the pump a few minutes while I clean the strainer."
- Fill caps are left off the tank.

These situations occur in most shops. It is economically impractical to control these activities. Repair work is done on the machine, on portable workbenches, or in maintenance shops. All of these areas tend to be dirty by nature. Management cannot hope to provide working areas and tools that are "clean" enough to prevent contamination during repair and maintenance activities. The solution to this apparent dilemma is to make the importance of system cleanliness known to the maintenance personnel through training activities. Awareness of the sources, the effect of contamination on the system, and its control will develop habits that are conducive to minimizing the introduction of dirt by shop activities.

Effects of Contamination on a Hydraulic System

It is well known that contaminant particles are of all shapes and sizes, and that the finer they are the more difficult it is to count them and to determine the material of which they are composed. However, we can say that the majority are abrasive and that when interacting with surfaces, they plough and cut little pieces from the surface. This wear accounts for about 90% of the failures due to contamination or dirt. The effect of these contaminant particles on various system components is different depending on the mechanism of operation. Following is a list of what such particles do to hydraulic components:

Pumps

- Erodes wear plates.
- Causes sticking of vanes, creating erratic action.
- Causes the vanes to wear out the cam ring.
- Wears out rotor slots.
- Increases shaft-journal and bearing wear.
- Increases gear wear with resultant inefficiency.
- In compensator controls, it causes sticking, slow response, and erratic delivery.
- Creates excessive heat and inefficient use of horsepower.

Relief Valves

- Dirt causes chatter.
- Accumulated dirt causes relief valve to fail safe, pressure becomes erratic.
- Dirt causes seat wear.

Directional Valves

- Dirt causes plugged orifices.
- Dirt causes wear to spool and housing lands, creating excess leakage.
- Dirt deposits cause spools to stick, which can cause solenoid failure.
- Sticking valves can cause excessive shock loads, damaging hose, piping, fitting, and other components.

Check Valves

- Dirt permits fluid to bypass check valve.
- Dirt causes wear on the ball and seats creating leakage.

Flow-Control Valves

■ Dirt causes erosion of orifices—changes the flow-setting characteristics.

Cylinders

■ Dirt causes excessive wear of the cylinder rod, piston seals, rod seals, and the tube bore.
■ Dirt causes cushions to malfunction.

Fluid Motors

■ Dirt causes wear similar to pump wear.

Fluid

■ Dirt acts as a catalyst, breaking down the molecular structure of the oil and causing a gummy residue to form—varnish.
■ Dirt in the tank tends to attract additives, which changes the composition of the fluid.

From the foregoing, it can be seen that failures arising from dirt or contamination can be classified into three categories:

1. **Catastrophic failure,** which occurs when a large particle enters a pump or valve. For instance, if a particle causes a vane to jam in a rotor slot, the result may well be complete seizure of the pump or motor. In a spool-type valve, a large particle trapped at the right place can stop a spool closing completely.
2. **Intermittent failure** is caused by contaminant on the seat of a poppet valve, which prevents it from reseating properly. If the seat is too hard to allow the particle to be embedded into it, the particle may be washed away when the valve is opened again. Later, another particle may prevent complete closure only to be washed away when the valve opens. Thus, a very annoying type of intermittent failure occurs.
3. **Degradation failure** follows wear, corrosion, and cavitation erosion. They cause increased internal leakage in the system components, but this condition is often difficult to detect.

Control of Contamination in a Hydraulic System

Nearly any discussion regarding dirt in a hydraulic system resolves itself to a consideration of particle sizes. For the most part, these are smaller than a grain of salt, or 100 microns. Forty microns is about the smallest particle that can be seen with the naked eye. One micron is equal to 39 millionths of an inch; twenty-five microns equals one-thousandth of an inch. Thus, when a 60-mesh filter (238 microns) is considered for use in a system, the openings are twice as large as a grain of table salt (100 microns); this size filter could not be effective in removing small, harmful particles. Figure 4.2 shows the relative size of micronic particles along with linear dimension equivalents.

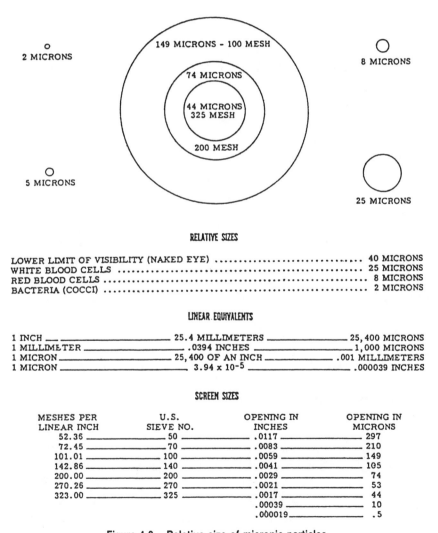

Figure 4.2. Relative size of micronic particles.

It is generally recommended that filtration to at least a 25-micron range be provided for a hydraulic system. This recommendation is made with respect to particle removal and is subject to the following qualifications:

- Filters located in a pump-suction line must be correctly sized to prevent cavitation.
- There are systems where better filtration is required because of very small clearances. In such systems, recommendations should be sought individually from component manufacturers.
- When filtering fire-resistant fluids, specific recommendations of the component and fluid manufacturers should be followed.

The critical clearances of the components used in the system are the determining factor for the degree of cleanliness required by the system. Table 4.1 shows critical clearances for some typical components.

Cleanliness level standards have been developed and accepted by industry. The three most commonly used and accepted are the IS0 Code, NAS 1638, and SAE Classifications. Table 4.2 indicates the fluid cleanliness required for various hydraulic components based on their critical clearances. Table 4.3 correlates the cleanliness levels for the various industry standards.

The maximum contamination levels permitted by NAS 1638 standard for the different class systems is shown on Table 4.4. A typical industrial hydraulic system is rated Class 7, NAS 1638, or 17/14 - 16/13, IS0 Code. Table 4.5 shows the particle counts obtained from two typical new hydraulic oil samples. It can be seen that both oil samples are well outside the needed cleanliness level for a typical system. Thus, the new oil requires filtration in order to meet the system specification requirements. This filtering can be done with the use of a portable filtration unit (Figure 4.3).

Some dirt particles in a system are magnetic. They are built into the system while the machinery is being fabricated. They are also generated within the system from the action of moving parts and fluid erosion. They can also enter the system through the reservoir openings and air breather.

These particles are normally abrasive and can react chemically with hydraulic oil to decompose it. They should be removed from the system. However, most of this type of dirt consists of very small particles. A fine filter would have to be used, which would mean an increase in cost and probably an increase in maintenance. Both of these factors can be avoided by using a magnet. If many magnetic particles are

Table 4.1
Typical Critical Clearances

Item	Typical Clearance	
	Microns	Inches
Gear Pump (Pressure Loaded)		
Gear to Side Plate	1/2–5	0.000,02–0.000,2
Gear Tip to Case	1/2–5	0.000,02–0.000,2
Vane Pump		
Tip of Vane	1/2–1*	0.000,02–0.000,04
Sides of Vane	5–13	0.000,2 –0.000,5
Piston Pump		
Piston to Bore (R)†	5–40	0.000,2 –0.001,5
Valve Plate to Cylinder	1/2–5	0.000,02–0.000,2
Servo-Valve		
Orifice	130–450	0.005 –0.018
Flapper Wall	19–63	0.000,7 –0.002,5
Spool-Sleeve (R)†	1–4	0.000,05–0.000,15
Control Valve		
Orifice	130–10,000	0.005 –0.40
Spool-Sleeve (R)†	1–23	0.000,05–0.000,90
Disc-Type	1/2–1*	0.000,02–0.000,04*
Poppet-Type	13–14	0.000,5 –0.001,5
Actuators	50–250	0.002 –0.010
Hydrostatic Bearings	1–25	0.000,05–0.001
Antifriction Bearings	*1/2–	0.000,02–
Slide Bearings	*1/2–	0.000,02–

* *Estimate for thin lubricant film*
† *Radial Clearance*

present in a system, a relatively coarse element used in conjunction with a magnet can be as effective as a finer filter. The magnet will catch the small metal particles. The element will catch the larger dirt and will not become clogged as quickly as a finer element. Cost and maintenance are reduced (Figure 4.4).

It is especially recommended that magnets be used with fire-resistant fluids. Petroleum oil allows many of the metal particles to settle to

Table 4.2
Fluid Cleanliness Required for Typical Hydraulic Components

Required for Typical Hydraulic Components	
Components	Fluid Classification ISO Code
Servo Control Valves	14/11
Vane and Piston Pumps/Motors	16/13
Directional and Pressure Control Valves	16/13
Gear Pumps/Motors	17/14
Flow Control Valves, Cylinders	18/15
Aircraft Test Stands	13/10
Injection Molding	16/13
Metal Working	17/14–16/13
Mobile	18/15–16/13
New Unused Oil	18/15

the bottom of the reservoir. Fire-resistant fluids are more detergent and tend to keep these particles in suspension. Consequently, there can be more magnetic particles in a stream of fire-resistant fluid than in a stream of petroleum oil.

Filter Elements

Filter elements are divided into two general classifications, depth-type filter elements and surface-type filter elements.

Depth-type filter elements force the fluid to pass through multiple layers of material. The dirt is caught because of the intertwining path that the fluid takes. These are generally absorbent-type elements since the dirt particles are trapped by the walls. An absorbent element causes the dirt to stick to the surface.

Because of its construction, a depth-type filter element has many pores of various sizes. Since there is no one consistent pore size, the element is generally given a nominal rating that is based on its average pore size. For example, an element with a nominal rating of 25 microns means that the average pore size is at least 25 microns; initially it will remove most contaminants 25 microns and smaller in size.

(text continued on page 184)

Table 4.3
Cleanliness Level Correlation

ISO Code	Particles/Milliliter		NAS 1638 (1964)	"SAE" Level (1963)
	≥5 Micrometers	≥15 Micrometers		
26/23	640,000	80,000	—	—
25/23	320,000	80,000	—	—
23/20	80,000	10,000	—	—
21/18	20,000	2,500	12	—
20/18	10,000	2,500	—	—
20/17	10,000	1,300	11	—
20/16	10,000	640	—	—
19/16	5,000	640	10	—
18/15	2,500	320	9	6
17/14	1,300	160	8	5
16/13	640	80	7	4
15/12	320	40	6	3
14/12	160	40	—	—
14/11	160	20	5	2
13/10	80	10	4	1
12/9	40	5	3	0
11/8	20	2.5	2	—
10/8	10	2.5	—	—
10/7	10	1.3	1	—
10/6	10	.64	—	—

Table 4.4
Maximum Contamination Levels Per 100 Milliliters

Particle sizes in microns	Particle count per 100 ml						
	Class 1	Class 2	Class 3	Class 4	Class 5	Class 6	Class 7
5– 15	500	1,000	2,000	4,000	8,000	16,000	32,000
15– 25	89	178	356	712	1,425	2,850	5,700
25– 50	16	32	63	126	253	506	1,012
50–100	3	6	11	22	45	90	180
100+	1	1	2	4	8	16	32

* *NAS 1638 Standard*

Table 4.5
Particle Count For Two Typical New Oil Samples

Particle sizes in microns	5–14	15–24	25–49	50–100	Over 100
Number of particles per 100 ml	135,000 520,000	46,000 150,000	18,600 50,000	5,400 14,000	4,700 12,000

For interest this table shows a typical particle count for two samples of new oil, from which it can be seen that this cannot be considered clean oil if it is to be used in a system

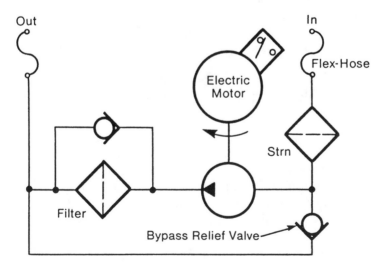

Figure 4.3. Hydraulic oil filtration cart. (Courtesy of Vickers Division, Trinova Corporation.)

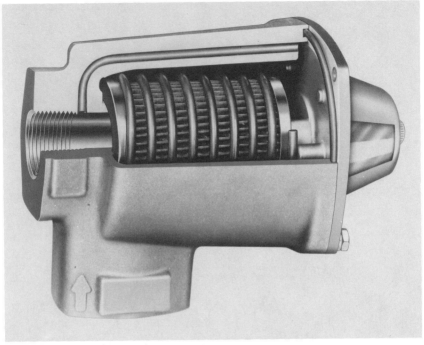

Figure 4.4. Inlet-line filter with magnet and condition indicator. (Courtesy of Vickers Division, Trinova Corporation.)

(text continued from page 180)

Advantages of depth elements:

- High efficiency on a one-time fluid pass basis
- Large dirt-holding capacity
- Usually inexpensive

Disadvantages of depth elements:

- Impractical to clean
- Limited compatibility with fluids
- High initial clean-pressure differential
- Paper and synthetic-type depth elements are adversely affected by temperatures above 260°F
- Paper elements have limited shelf life
- Paper-type elements are not recommended for use with water-based fluids
- Paper-type elements are impractical to use with fluid viscosities above 1000 SUS

Surface-type filter elements consist of a single layer of material through which the fluid must pass. The material layer consists of perforated metal or woven wire-mesh screening. The manufacturing process of these material layers can be very precisely controlled so that these elements can have very consistent pore or opening sizes. Because of this, surface-type elements can be identified by their absolute rating. The "absolute" rating signifies the largest opening in the filter element. Thus, this rating indicates the largest, hard, spherically-shaped particle that can pass through the filter.

Advantages of surface elements:

- Strength and resistance to fatigue, temperature, and corrosion
- Surface-type elements can be cleaned and reused
- Maximum pore size can be controlled
- Low intitial clean-pressure differential

Disadvantages of surface elements:

- Usually expensive
- Not as efficient as depth-type element on a one-time fluid-pass basis

Filter Location

In addition to the degree of filtration, the location of the filters within a system is an important consideration. Hydraulic filters can be installed in the intake, pressure, or return lines of the system or in the reservoir (Figure 4.5).

Intake-line filters prevent catastrophic failure of pumps by collecting "large" contaminant particles. The filters are usually mounted either partly or completely outside the system reservoir, near the pump intake. They are, most often, relatively coarse, which makes them ineffective in controlling dirt levels.

Most designs provide for servicing of the filter element through the filter housing without opening or draining the system reservoir. Some designs even allow servicing without shutting down the system.

Intake filters must be sized carefully to assure that the pump can tolerate pressure drop at maximum flow and minimum operating temperature, assuming a plugged filter element. Otherwise, the pump may "starve" and cavitate.

Pressure-line filters are used on the outlet side of the pump or just ahead of valves and other highly dirt-sensitive components such as

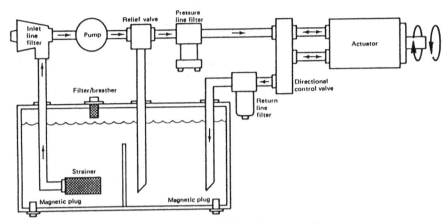

Figure 4.5. Typical locations of filters in a hydraulic system.

servovalves and proportional valves. Pressure-line filters remove critical contaminants passing through or generated by the pump before they get into the remainder of the system. Any contamination generated in the system downstream of the pressure-line filter will increase dirt levels in the reservoir unless controlled by a return-line filter. The same is also true for dirt that enters the system through cylinder-rod seals.

Return-line filters are normally placed in the system return line to clean the hydraulic fluid before it enters the reservoir. Return-line filtration is especially recommended for start-ups of new systems because most initial contamination comes from within the system itself— contaminants inadvertently left in piping and components, metal chips from manufacturing and assembly, moving surfaces and threaded fittings, and particles from thread-sealing materials, hoses, elastomer seals, etc.

Each filter location offers some advantages over any other. For example:

- Only a return-line filter can reduce the amount of dirt entering the system through a cylinder-rod seal.
- Only an intake filter can prevent a large particle that may be in the reservoir from entering the pump and possibly causing a catastrophic failure.
- Only a pressure-line filter at the pump outlet can prevent particles from a wearing or failing pump to be spread throughout the system. These particles could cause valves, cylinders, and motors to fail.

All locations are important; however, economics generally dictates that all three filters cannot be placed in the system. The use of only one or two filters in a system requires the careful analysis of the dirt-level tolerances of the components and the consequences of a failure. The analysis should also include safety, potential machine damage, maintenance, and repair costs.

Filter Maintenance

A hydraulic system may be equipped with the best filters available and they may be positioned in the system where they do the most good; however, if the filters are not taken care of and cleaned when dirty, the money spent for the filters and their installation has been wasted. Thus, the whole key to good filtration is filter maintenance. Following are some suggestions that will help in providing proper filter maintenance:

1. A filter-maintenance schedule should be set up and followed diligently.
2. Inspect filter elements that have been removed from the system for signs of failure and of impending system problems.
3. Any fluid that has leaked out should not be returned to the system.
4. The supply of fresh fluid should be kept covered tightly.
5. Use clean containers, hoses, and funnels when filling the reservoir.
6. Use commonsense precautions to prevent entry of dirt into components that have been temporarily removed from the circuit.
7. All clean-out holes, filler caps, and breather-cap filters on the reservoir should be properly fastened.
8. Do not run the system unless all normally provided filtration devices are in place.
9. Make certain that the fluid used in the system is of a type recommended by the manufacturers of the system or components.
10. Before changing from one type of fluid to another (for example, from petroleum-based oil to a fire-resistant fluid), consult component and filter manufacturers in selection of the fluid and the filters that should be used.

Filter Bypass Valve

If filter maintenance is not performed, pressure differential across a filter element will increase (Figure 4.6). An unlimited increase in pressure differential across a filter on the pressure side means that the

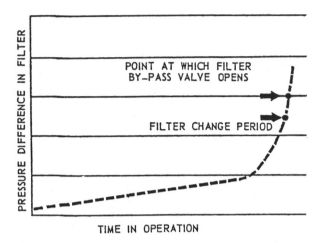

Figure 4.6. Life of a filter element.

filter element will eventually collapse or dirt may be pushed through. To avoid this situation, a simple relief valve is used to limit the pressure differential across a full-flow filter. This type relief valve is generally called a bypass valve.

There are several types of bypass valves, but they all operate by sensing the difference in pressure between dirty and clean fluid. A bypass valve basically consists of a movable piston, housing, and spring, which biases the piston (Figure 4.7).

Pressure of dirty fluid coming into the filter is sensed at the bottom of the piston. Pressure of the fluid after it has passed through the filter element is sensed at the other side of the piston on which the spring acts. As the filter element collects dirt, the pressure required to push the dirty fluid through the element increases. Fluid pressure after it passes through the element remains the same. When the pressure differential across the filter element, as well as across the piston, is large enough to overcome the force of the spring, the piston will move up and offer the fluid a path around the element.

A bypass valve is a fail-safe device. In an inlet filter, a bypass limits the maximum pressure differential across the filter if it is not cleaned. This protects the pump. If a pressure or return-line filter is not cleaned, a bypass will limit the maximum pressure differential so that dirt is not pushed through the element or that the element is not collapsed. In this way the bypass protects the filter.

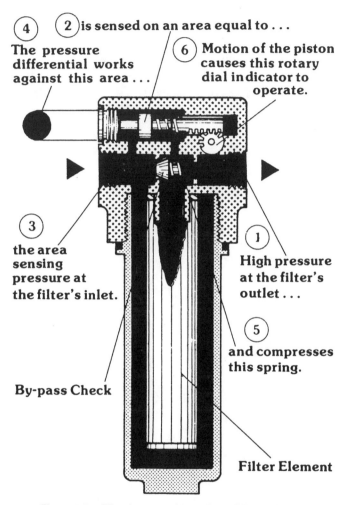

④ ②is sensed on an area equal to . . .

The pressure differential works against this area . . .

⑥ **Motion of the piston causes this rotary dial indicator to operate.**

③

the area sensing pressure at the filter's inlet.

①

High pressure at the filter's outlet . . .

⑤

and compresses this spring.

By-pass Check

Filter Element

Figure 4.7. Filter bypass valve and condition indicator.

Filter-Condition Indicator

The bypass valve protected the filter by offering the fluid an alternate path when the pressure difference across the filter element exceeded some safe value. When the bypass valve is open, the filter is in effect out of the circuit and does not effect any dirt control. This condition is not tolerable for any length of time since system degradation will take place. The remedy is to clean, replace, and maintain the filter whenever such a condition is encountered. To help in this regard, a filter can be equipped with a condition indicator. This item indicates when the element is clean, needs cleaning, or is in the bypass condition. Indicators are available that are dependent on the motion of the bypass piston; also, units that sense the pressure differential across the filter element are offered by several manufacturers.

Compressed-Air Filtration, Lubrication, and Moisture Control

Contaminants in the Air System

Dirt contaminates a pneumatic system. It is very similar to industrial smokestacks spewing soot. However, dirt is only one form of contamination in a pneumatic system. To determine what other elements harm a pneumatic system and how they are handled, let us follow the step-by-step passage of air through a pneumatic system. A logical place to start our journey is with the industrial air made available to the compressor (Figure 5.1).

In an industrial environment, air carries several undesirable elements as far as a pneumatic system is concerned. Two of these are dust and water vapor. Depending on the type of system and the degree of sophistication, dust and water vapor must be removed either partially or totally if a degree of system dependability is to be realized. A compressor is sometimes considered the heart of a pneumatic system. If the compressor fails, the system cannot function.

Air is often described as a mixture of gases made up of 21% oxygen, 78% nitrogen, and 1% other ingredients. Air is colorless, odorless and tasteless. Identifying air by its physical properties is slightly misleading because in a pneumatic system air is not dealt with in this pure form. Besides being all the things described above, air is also a carrier.

Pneumatic-system pollution is measured using the micrometer scale. One micrometer (μm) is equal to one millionth of a meter, or 39 millionths of an inch. A single micrometer is invisible to the naked eye and is so small that even to imagine it is extremely difficult. To bring the size more down to earth, two everyday objects measured using the micrometer scale have the following values: an ordinary grain of table salt measures 100 μm, and the average diameter of human hair

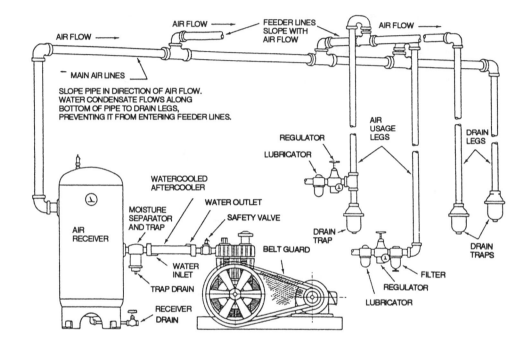

Figure 5.1. Air compression and conditioning system.

measures 70 μm. Twenty-five μm is approximately .001 in. The lower limit of visibility for an unaided human eye is 40 μm. In other words, the average person can see dirt that measures 40 μm and larger. Some of the harmful dirt particles of a pneumatic system are smaller than 40 μm.

A compressor is protected by an intake filter, which is engineered into the inlet by the compressor manufacturer. It is designed to remove dust particles commonly found in an industrial environment. Industrial dust particles are not necessarily the same type of dust you might find on a piece of furniture in your living room. Depending on the type of manufacturing being done, industrial dust can be iron, carbon, grinding wheel particles, silicates, fiberglass, and abrasive materials of all kinds. This pollution must be prevented from entering a compressor or it will harm compressor seals as well as interfere with the operation of downstream system components such as regulators, directional-control valves, flow-control valves, and cylinders.

An intake filter is a very important part of a pneumatic system. It is the first line of defense against industrial air. For this reason, intake filters should be serviced at regular intervals to achieve optimum dependability.

After being filtered to a large extent by the compressor-intake filter, air moves to the piston chambers. If it is a two-stage compressor, air is compressed, with a consequent temperature rise in the first stage; passed through an intercooler, where its temperature is decreased; and then moves on to the second stage. In the second stage, the air is compressed more and has another increase in temperature. It is then pushed into the system.

The air at this point has increased in potential energy because it is compressed. This is desirable, but the air is now hot and contains water vapor. It may contain some dirt that was not removed by the intake filter, and it may carry some oil vapor that was picked up while passing through the compressor. This is not desirable. Instead of discharging the air from the compressor outlet directly into an air receiver for storage, the air is often passed through an aftercooler. As its name implies, an aftercooler cools compressed air. This is accomplished by passing cooling water or air over the aftercooler chamber. Besides being the point where air cools, an aftercooler is also the place where some dirt and oil vapor fall out of suspension, and a good portion of entrained water vapor condenses.

Moisture can affect the operation of a pneumatic system in several ways:

1. Moisture can wash away lubricant from pneumatic components, resulting in faulty operation, corrosion, and excessive wear.
2. Moisture that collects in pneumatic cylinders can cause cushions to become ineffective.
3. Moisture that condenses at work outlets can cause an operation or final product to be ruined. It is also an annoyance and inconvenience to the operator.

As air travels through piping, it picks up pieces of dirt, such as rust and pipe scale. It will also contain any moisture that was not removed in the aftercooler. Before air reaches a directional valve and actuator downstream, it must be filtered again. This is the function of an air-line filter.

Air-Line Filters, Lubricators, and Pressure Regulators

Compressed Air Filters

Filtration is the first, and often the only, line of defense against contamination in a pneumatic system. Adequate filtration is seldom expensive and will return the investment many times over in improved component life and reduced downtime.

Contaminants

Compressed air contains all the dust, pollen, and other solid impurities that were in the free air that entered the compressor inlet, but with one major difference: At 100 psi, the concentration of contaminants is nearly eight times greater than it was in free air. In addition, the compressed air can also pick up rust, scale, and other dirt particles from the steel piping, which carries it to the point of use. Excess pipe sealant (from careless plumbing) is another contaminant. Other construction debris are bound to cause problems long after installation is completed (Table 5.1).

In many cases, even the compressor contributes metal fragments or fine solid particles of carbon or Teflon from sliding seals. If there is a desiccant dryer in the system, desiccant particles may be carried downstream.

Air leaving the compressor is usually not only saturated with water vapor, but has probably picked up oil vapor from the compressor. As air moves through the cooler areas of the whole pneumatic system, some of these vapors condense as liquid particles, which are carried along in the airstream. For predictable operation and reliable service life, both solid and liquid particles must be removed by the pneumatic filters.

Filter Operation

A typical industrial air filter should technically be described as a combination of a dynamic separator and a mechanical or static filter. An air-line filter consists basically of a housing with inlet and outlet ports, deflector plate, filter element, baffle plate, filter bowl, and drain

Table 5.1
Typical Problems Caused by Wet, Dirty Air

Application	Problem
Pneumatic instruments and controls	Corrosion of internal parts and deterioration of seals, resulting in malfunction
Outside air lines	Freezing, line blockage, rupture
Chemical blanketing	Corrosion of containers, piping, and valves
Pneumatic tools	Rust and pitting of internal parts
Sand or grit blasting	Clogging of feed lines and spray nozzles
Material handling	Wet cakes, sticking, and plugged lines
Spraying—paints and coating	Surface blemishes
Air bearings	Corrosion

(Figure 5.2). Air entering the inlet is deflected by the deflector plate. This causes a swirling action, which throws out large pieces of dirt and liquid droplets against the wall of the bowl. When the swirling air hits the baffle, its direction is again changed. It then passes through a filter element, where fine pieces of dirt are removed. It then exits the outlet port.

The area below the filter baffle is known as the "quiet zone". This is the place where large contaminants and liquid collect. Often, filter bowls are made of a transparent plastic material so that buildup in the bowl can be observed. When the collection of contaminant and liquid approaches the baffle, the drain cock in the bottom of the bowl should be opened. Some air-line filters are equipped with an automatic drain.

After separation of large solid and liquid particles, the compressed air is directed through a filter element that strains out smaller particles, subject to the size limitations of the element's design. Air flows from the outside to the inside of the element.

Filter elements used in air-line filters are divided into depth and edge types. *Depth-type* elements force air to pass through an appreciable material thickness. Dirt is trapped because of the intertwining path

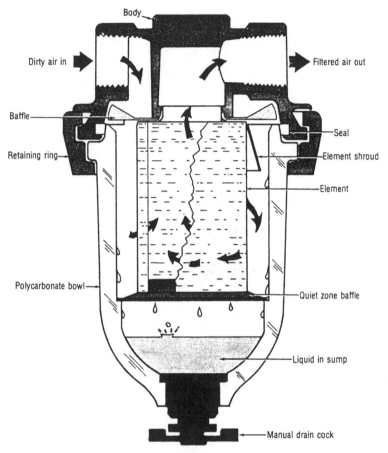

Figure 5.2 Air-line filter assembly.

the air must take. Depth-type elements in pneumatic systems are frequently made of porous bronze. *Edge-type* elements offer an airstream with a relatively straight flow path. Dirt is caught on the surface or edge of the element that faces the air flow. Edge-type elements in pneumatic systems are often made of resin-impregnated paper ribbon.

Because of its construction, an air-line filter element has many pores of various sizes. Many of the pores are small. A few pores are relatively large. Since it has not one consistent hole or pore size, an air-line filter element is given a "nominal rating." A nominal rating is an element rating given by the filter manufacturer. For example, a depth element with a nominal rating of 40 μm indicates that on the average, the pores in the element are 40 μm in size.

Air-line filter elements found in industrial pneumatic systems generally have nominal ratings ranging from 50 to 5 μm. Since dirt in a pneumatic system comes in all sizes, shapes, and materials, no guarantee is made by a nominal rating as to what size particles will be removed from a compressed airstream.

Drains

There are many manual and automatic drains to remove collected liquids from filter bowls. Miniature filters, where size is a considera-tion, use a tire valve type of drain with an external pushpin to act as an actuator. Larger filters are designed with metal or plastic drain cocks that are actuated by wing nuts or knobs. Some manufacturers equip their manual drains with a flexure tube which opens the drain whenever the tube is pushed away from its normal, vertical position (Figure 5.3).

The automatic filter drain (Figure 5.4) is a valuable option when filter location makes servicing difficult or dangerous, or when filters are hidden and may be inadvertently overlooked. The two basic types of automatic drains are pulse-operated for intermittent air flow and float-operated for continuous flow. In the *pulse-operated* drain, the drain valve is linked mechanically to a piston or diaphragm that fits across the bowl above the surface of the sump. When fluid pressure is equal on both sides of the piston, the larger area holds the drain valve seated. As flow starts through the filter, the pressure drops temporarily above the piston; the piston unseats the drain valve, emptying the sump. Pressure above the piston then reseats the drain valve. Flow through the filter must be intermittent for the pulse-operated drain to work.

In *float-operated* drains, accumulated liquids in the sump raise a float that either opens the drain mechanically, or pilots a valve that opens the drain. Bowl pressure forces the liquid out rapidly, the float lowers, and the drain closes. The drain can be mounted internally or externally as shown on Figure. 5.5.

Types of Elements

Industrial air filter elements are available in a variety of materials, including felt, paper, cellulose, metal and plastic screening, metal rib-bon, sintered bronze, sintered plastic, glass fiber, and cloth. Elements

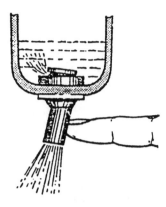

Figure 5.3.   Manual drain on an air-line filter.

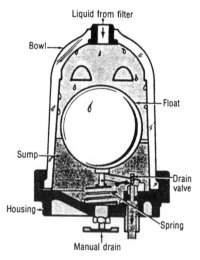

Figure 5.4.   Float-type automatic drain for air filter.

are rated according to the minimum-size particle they will remove from the airstreams. Elements are also rated for flow through the filter at specific pressure. Some manufacturers classify elements as edge or depth type. Elements are also listed as cleanable and disposable. Edge-type elements are easy to clean; depth-type elements are more difficult to clean.

INTERNAL AUTOMATIC DRAIN

EXTERNAL AUTOMATIC DRAIN

Figure 5.5. Internal and external automatic drains.

Coalescing describes an act of growing together. Coalescing elements for oil aerosol removal starts with a random bed of borosilicate micro-fibers. To make the fibers self-supporting, they can be bonded with resin, or another, stiffer fiber can be mixed with the borosilicate. Another construction method supports the borosilicate fibers mechanically with external layers of strong material. A sleeve over the outer surface of a coalescing element helps prevent the airstream from reentraining liquid oil from the wet surface of the element. Adsorption is described as adhesion of the molecules of a gas, liquid, or dissolved substance to a surface. Adsorption of oil vapors takes place in either packed columns of activated carbon (the adsorptive medium) or in elements that have activated carbon particles mixed with filtering fibers.

Filter Applications

Certain general considerations apply to all air-line filter units. Most important, units should be sized according to airflow demands, not according to pipe size. For instance, an oversized filter may not impart enough swirling action to separate all the moisture droplets. Conversely, an undersized filter creates too much turbulence in the air being processed. The turbulence prevents the liquids from settling into the quiet zone. Undersized units may also add excess pressure loss to the system. Flow capacities for a given port size also vary among manufacturers because of variations in internal construction.

Air-drop lines to conditioning units should be attached to the top of the main header to decrease the amount of liquid that might run down to equipment. A separate drain should be provided for those liquids that accumulate at low points in the main header line.

Conditioning units should be installed as close as possible to the point of use, consistent with accessibility for inspection and possible servicing. Bends in the air lines should be avoided since they cause additional pressure drop in the system. In addition, momentum forces tend to deposit lubricant at bends. Necessary turns should have as large a radius as possible.

Filters should be installed upstream from, and as close as possible to, the devices to be protected. Filters remove only droplets and solid contaminants; water vapor is not affected. Thus vapor can condense between the filter and the components being served. To minimize this possibility, filters should be located as close as possible to the point of use. The filter bowl must hang vertically so that free moisture and oil droplets cling to the inner surface of the bowl as they work their way down to the quiet zone at the bottom of the bowl. Filters should be located so that manually drained bowls can be observed and drained when necessary. Drain lines should be provided for filters that incorporate automatic drain devices.

Normally, filters have no moving parts to service or adjust. However, the filter element should be replaced when the pressure differential across the filter unit exceeds 10 psi. Although a manual drain requires regular attention, it is necessary to drain the collected liquids and solids only when their level approaches the lower baffle.

Lubricators

After compressed air passes through an air-line filter, it is clean and relatively free of entrained water. But for some applications, the air must be conditioned in still one more way—it must be lubricated. Lubrication of compressed air may be necessary to provide seal lubrication, prevent sticking of moving parts, and control wear. Air-line lubricators work on the basis of developing a controlled mist of lubricant in an airstream (Figure 5.6).

Air entering the inlet runs into a constriction. This results in a higher pressure at the inlet to the constriction with respect to the small opening of the mist chamber. This pressure differential is sensed across the pickup tube, pushing fluid up the tube and into the mist chamber, where it is mixed with air. The air then exits the outlet carrying suspended oil particles. The amount of oil mist that exits from this type of lubricator depends on the adjustment of the needle valve.

The *mist lubricator,* the most commonly used in-line lubricator, is of the direct-flow type. It permits every drop of oil in the sight dome to go downstream into the air line. All the oil observed in the sight dome flows downstream in aerosol form, the particles ranging in size from 0.01 to 500 μm. This lubricator can be filled while the air line is under pressure because it contains a pressurization check valve.

The *micromist lubricator* was developed to fulfill the requirements of more difficult applications. It is a recirculating-flow-type lubricator. This means that the oil, once injected into the air, is recirculated into the bowl of the lubricator, where the larger oil particles fall out of the air. Only the smaller oil particles that float around in the upper part of the bowl go downstream. The particles going downstream in this type of lubricator range in size from 0.01 to 2 μm. This lubricator,

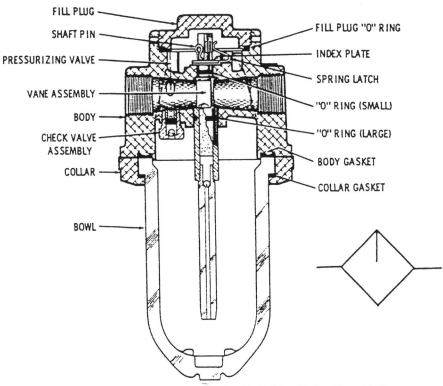

Figure 5.6. Air-line lubricator. (Courtesy of Schrader Division, Parker-Hannifin Corporation.)

because of its design, cannot be filled under pressure unless a remote fill button is incorporated into the lubricator.

Oil mist flowing from the lubricators will remain suspended over varying distances. Direct-flow mist lubricators can achieve oil particle suspension in the line over a distance of 20 ft, utilizing the greater portion of the oil supply available. By comparison, the recirculating-flow micromist lubricator can keep the majority of the oil suspended in the piping for approximately 100 ft. This can be achieved because of the absence of the larger oil particles in the output of the micromist lubricator. The larger particles in the mist lubricator tend to "eat up" or "coalesce" the smaller particles as that aerosol mixture flows downstream. Because of the absence of these larger particles in the micromist lubricator, this "coalescing" or "eating up" effect is not prevalent. Consequently, the oil particles that do come from the lubricator can remain suspended over considerable longer distances of piping.

Micromist lubricators are particularly well suited for the following applications:

1. Where instant oil delivery and lubrication are required, whether the air line is wet or dry.
2. In systems where multipoint lubrication is required, such as lines that contain several valves and cylinders.
3. Where large oil drops from the air exhaust would contaminate either the environment or the material being processed (the micromist lubricator actually permits less oil to be injected into the air line to perform the lubrication job).
4. Where the air line includes (after the lubricator) valves, bends, and similar baffling points that would tend to separate out heavier particles.
5. Where ease of drip-rate adjustment is required.
6. For high-cycle pneumatic applications.
7. Where very low flow rates are necessary.
8. Where the oil particles stay in suspension in the air lines for 100 ft or more.
9. Where the oil particles must stay in suspension for more than 10 minutes.
10. Where the oil mist becomes a vaporous part of the air and travels as part of the air until it is baffled out by bearings, vanes, and turbulence.

Refer to the manufacturer's instruction manual as to the method of adjusting oil flow and safety precautions while servicing the lubricator.

Remember that oil injected into an air stream is eventually blown out as an exhaust. If too much is blown into a confined space, the oil content of the air may become greater than allowed by regulations.

A pneumatic system may be equipped with the best filters and lubricators available, and they may be positioned in the system where they do the most good; however, if they are not serviced when required, the money spent for their use has been wasted. Lack of proper lubrication in pneumatic power components creates excessive maintenance costs, production inefficiency, and premature failure. Proper lubrication allows components to operate with a minimum of friction and corrosion, and it minimizes wear on the component parts. The most economical way to lubricate pneumatic components is with an air-line lubricator. In most cases it is also the most efficient.

Pressure Regulators

The amount of pressure needed in a system depends on the force required to do a job. Because some applications require less force than others, it follows that the required pressure varies accordingly. It is also essential that once a system pressure has been selected, air be supplied at constant pressure to each tool or actuator regardless of variations in flow and upstream pressure.

Thus, it is important to add to the system a pressure regulator that performs the following functions:

- Supplies air at constant pressure regardless of flow variation or upstream pressure.
- Helps operate the system more economically by minimizing the amount of pressurized air wasted. This happens when the system operates at pressures higher than those needed for the job.
- Helps promote safety by operating the tool or actuator at a reduced pressure.
- Extends component life because operating at higher-than-recommended pressures increases the wear rate and reduces equipment life.
- Increases operating efficiency.

Types of Regulators

Unbalanced poppet, non-pilot-operated type. Figure 5.7 shows an unbalanced poppet regulator, the simplest pressure regulator. Supply pressure enters the inlet port and flows around the poppet. However, as shown, the poppet is seated on the orifice, not permitting flow. Turning the adjustment screw down compresses the adjustment spring, forcing the diaphragm downward. The diaphragm pushes the stem down and the poppet moves away from the orifice. As downstream pressure builds, it acts on the underside of the diaphragm, balancing the force of the spring, and the poppet throttles the orifice to restrict flow and produces the desired pressure. As the demand for downstream flow varies, the regulator compensates by automatically repositioning the poppet in relation to the orifice. The bottom spring of the poppet assures primarily that the regulator will close at no-flow.

Unbalanced poppet, non-pilot-operated, with diaphragm chamber. The next most complex regulator is shown in Figure 5.8. It is also an unbalanced poppet, non-pilot-operated type. However, it has a

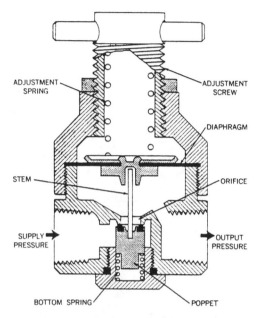

ADJUSTMENT
SPRING

ADJUSTMENT
SCREW

DIAPHRAGM

STEM

ORIFICE

SUPPLY
PRESSURE

OUTPUT
PRESSURE

BOTTOM SPRING

POPPET

Figure 5.7. Unbalanced poppet, nonrelieving air-line pressure regulator.

diaphragm chamber that isolates the diaphragm from the main airflow. This reduces the abrasive effect of the air on the diaphragm.

The diaphragm chamber is connected to the outlet chamber through an aspirator tube. This tube causes a slightly lower-than-outlet pressure in the diaphragm chamber as flow increases through the regulator. This reduces the pressure drop experienced when the regulator opens.

Remote-controlled, pilot-operated balanced poppet. With some applications, the regulator must be installed where it cannot be easily adjusted. In such cases, the regulation and pressure-setting mechanisms are separated. A small air-pilot line connects the regulator (in the air line at the point of use) to the remote-setting mechanism, where it is convenient to make the adjustment, such as a control panel (Figure 5.9). The remote-setting mechanism is a small regulator used to produce a control-air signal.

Piston-type regulators. Although diaphragm-type regulators have been used to illustrate these basic types of regulators, almost every example is available as a piston regulator, where a piston replaces the diaphragms.

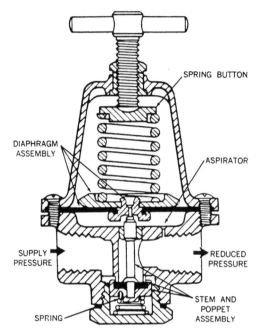

Figure 5.8. Unbalanced poppet, self-relieving air-line pressure regulator.

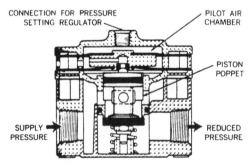

Figure 5.9. Remote-controlled, pilot-operated pressure regulator.

Air-Line Pressure-Regulator Applications

A regulator should be located for convenient servicing. The pressure setting should be adjusted to the value required by typical operating conditions. The no-flow pressure setting should not differ from the secondary pressure under flow conditions. A slight decrease in secondary pressure with increasing flow through the unit is normal.

Regulator flow characteristics are affected by the spring rate used to cover the desired pressure range. Generally, a standard unit is rated for 250 psig primary supply pressure, and adjustable secondary pressure is rated from 3 to 125 psig. Low-pressure and high-pressure units are also available.

If an application requires 30 psig, the standard unit will probably be satisfactory. If optimum performance is required, the low-pressure unit can be used, since its lower-rated spring gives better performance characteristics. The standard, adjustable secondary pressure range is satisfactory for most industrial pneumatic applications.

All secondary pressure requirements are covered by one pilot-controlled regulator, since the secondary pressure range of a pilot-controlled regulator is determined by the range of its pilot regulator. If the pressure and flow requirements of a system vary widely, a pilot-controlled regulator with a high-pressure pilot regulator provides the best performance. If the decrease in pressure under flow conditions seems excessively high for a particular application, the supply lines, the regulator unit, or other system components may be undersized for the application. A clogged filter element can also cause an excessive decrease in pressure under flow conditions.

To lower a setting, the regulator should be reset from a pressure below the final secondary pressure desired. For example, to lower the secondary pressure from 80 psig to 60 psig, decrease the secondary pressure to 55 psig or less, then adjust it upward to 60 psig.

In a nonventing regulator, some secondary air must be exhausted to lower the setting. Many regulators include a constant bleed of secondary air to obtain good pressure-cracking characteristics. A small fixed bleed in the secondary line provides the same effect. This modification is especially important in no-flow conditions, and permits the regulator to provide immediate response to variations in settings.

Pilot-controlled regulators are applied in the same way as standard spring-controlled regulators. The distance between the pilot-controlled regulators and its pilot regulator is not critical. Regulators do not require routine maintenance in normal service when properly protected by filters. The secondary pressure setting should be checked, however, whenever system requirements change. The combination of an air line filter, pressure regulator, and lubricator is commonly called an F-R-L or "trio unit." A typical such unit is shown in Figure 5.10.

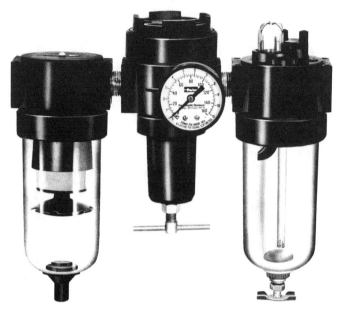

Figure 5.10. Air-line filter, pressure regulator, and lubricator. (Courtesy of Schrader Division, Parker-Hannifin Corporation.)

Safety: Metal Bowls

Normally, clear plastic bowls perform with no difficulty, and they are highly desirable from the standpoint of visibility of condensate level or oil level (Figure 5.11). Metal bowls are available for most brands of F-R-L units. They are intended to provide a greater measure of safety when one of these unusual operating or environmental conditions is encountered:

1. *High temperatures.* If internal or external temperature will exceed 150°F, most plastic bowls lose their mechanical strength and should be replaced with metal bowls.
2. *Solvents.* If solvents are present either in the airstream or in the atmosphere surrounding the F-R-L, metal bowls should be used. The composition of plastic bowls varies some between manufacturers, and the literature should be checked to determine if a given solvent is harmful to the particular plastic bowl in use. Harmful conditions might include the use of a lubricator as a chemical injector in a gas line, a fire-resistant liquid mist in the airstream, or an antifreeze chemical in the lubricator to prevent moisture freeze-up in the lines.

CAUTION: These products are specifically designed for compressed air service. Use with or injection of certain hazardous fluids or gases in the system (for example, alcohol or liquified petroleum gases) could be harmful to the unit or result in a combustible condition or hazardous external leakage.

Beware of materials that will attack polycarbonate plastic bowls. When dirty replace bowl or clean with dry, clean cloth. A bowl guard should always be used with all polycarbonate plastic bowls.

THE FOLLOWING MATERIALS WILL HARM A POLYCARBONATE BOWL:

Acetaldehyde	Caustic potash solution (5%)	Ethylene dichloride	Phosphorous trichloride
Acetic acid (conc.)	Caustic soda solution (5%)	Formic acid (conc.)	Propionic acid
Acetone	Chlorobenzene	Freon (refrigerant & propellant)	Pyridine
Acrylonitrile	Chloroform	Gasoline (high aromatic)	Sodium sulfide
Ammonium fluoride	Cresol	Hilgard Co.'s hil-phene	Styrene
Ammonium sulfide	Cyclohexanol	Hydrochloric acid (conc.)	Sulfuric acid (conc.)
Benzene	Cyclohexanone	Methyl alcohol	Sulphural chloride
Benzoic acid	Cyclohexene	Methylene chloride	Tannergas
Benzyl alcohol	Dimethyl formamide	Milk of lime (CaOH)	Tetrahydronaphthalene
Bromobenzene	Dioxane	Nitric acid (conc.)	Thiophene
Butyric acid	Ethane tetrachloride	Nitrobenzene	Toluene
Carbolic acid	Ethyl ether	Nitrocellulose lacquer	Xylene
Carbon disulfide	Ethylamine	Phenol	Perchlorethylene and others
Carbon tetrachloride	Ethylene chlorohydrin	Phosphorous hydroxy chloride	

TRADE NAMES OF SOME SYNTHETIC COMPRESSOR OILS & RUBBER COMPOUNDS THAT HARM POLYCARBONATE BOWLS.
■ Cellulube #150 and #220 ■ Houghton & Co. oil #1120, #1130, and #1055 ■ Keystone penetrating oil #2 ■ Marvel Mystery Oil ■ Some Loctite Compounds ■ Sinclair oil "Lily White" ■ Haskel #568-023 ■ Parco # 3106 Neoprene ■ Sears Regular Motor Oil ■ Garlock #98403 (Polyurethane) ■ Kano Kroil ■ Pydraul AC ■ Stillman #SR 269-75 (Polyurethane) ■ Stillman #SR 513-70 (Neoprene) ■ Tenneco Anderol #495 and #500 oils

Figure 5.11. Cautions on the use of polycarbonate plastic bowls.

209

3. *High pressure.* Plastic bowls usually have a lower pressure rating than metal bowls; this is primarily because of the possibility of accidental exposure to solvents or high temperature. Under the usual atmospheric temperature conditions, most plastic bowls are as safe as metal bowls.

The pressure gauging port on the side of a pressure regulator is usually ⅛ or ¼ in. National Pipe Thread (NPT) and is connected into the regulator secondary. This is a convenient place to make another air takeoff for a small quantity of air. On some brands the gauge port has a full ¼-in. opening into the airstream and can be used for full capacity of a ¼-in. line. On other brands, while the gauge connection may be ¼-in. NPT, the actual passage inside the port may be only ¹/₁₆-in. in diameter. A visual inspection will determine its size.

Routine Maintenance of Filter-Regulator-Lubricator (F-R-L) Units

What Oil to Use

Air-line lubricators are designed to work on petroleum oil of low viscosity. The precise grade, viscosity, or brand is not extremely critical. For general use, a nondetergent motor oil of not greater than SAE 10W is acceptable. Avoid the use of oils with additives because of the possible effect of these additives on components that may be downstream.

How Much Lubrication

While oil-feed rate is not extremely critical, an effort should be made to get near the optimum amount. Too much oil can be detrimental. Experience has shown that a feed rate of 1 drop of oil to about every 20 standard cubic feet of air (free air) used is about right for most applications.

Excessive Oil Feed

Oil feed is excessive if a surplus drips out the exhausts of directional control valves. Reduce the lubricator feed rate and check again in a few days after the system has stabilized.

Cleaning Plastic Bowls

Periodically, the plastic filter bowl should be removed and cleaned. Use only a damp cloth, with a little soap if necessary. Do not use lacquer thinner, gasoline, or other solvents because of their possible harmful effect on the plastic. A bowl weakened by solvent may burst when under pressure. While cleaning the bowl, inspect and clean the filtering element. Replace it if necessary.

Troubleshooting

Air Leak at Regulator

A continuous air leak from the small vent hole in the regulator bonnet indicates a leaky main poppet or diaphragm. In either case repair parts should be ordered at once and the regulator scheduled for repair. Overhaul kits with diaphragm and seals are available for most standard regulators.

IMPORTANT! Replace diaphragm or seals as soon as possible after leak is discovered. A complete failure of the diaphragm might permit full inlet pressure downstream. Raising the pressure on some solenoid valves may cause them to shift by themselves, creating a safety hazard to personnel.

Pressure Regulator Problems

If adjusting the regulator fails to reduce the pressure downstream, check these possible causes:

1. The regulator may be connected backward, with inlet pressure connected to the regulator outlet. It positively will not work if connected in reverse.
2. The regulator may be connected improperly. A gauge port may have been accidentally used as one of the main ports. This is not a full-flow port on some regulators.
3. There may be broken parts inside the regulator. If the diaphragm is cracked or broken, a high-velocity air leak at the vent hole will pinpoint this trouble.
4. If the regulated pressure is too low and cannot be raised by screwing the adjustment all the way down, the range must be changed by replacing the main spring with a stronger one.

Filter Problems

Very few problems are encountered with air-line filters, provided that the element is cleaned or replaced regularly. Here are a few suggestions for best filtering action:

1. Make sure that the filter is correctly plumbed, with incoming air entering at the inlet. It will filter in the reverse direction, but the

centrifugal baffles will be useless, and large particles of contamination will collect on the inside of the element and cause severe air restriction, making the element difficult to clean.

2. If the filter manufacturer offers a choice, install a fine filtering element, a 5-μm rating being preferred. With present-day microfinishes on cylinder barrels, a finer degree of filtration becomes increasingly desirable.

3. Service the filter regularly, cleaning or replacing the element. If allowed to become overly contaminated, the pressure drop through the unit may increase to the extent that speed in the air circuit will be significantly reduced.

4. Drain the condensate before it rises above the baffle. If allowed to rise above the baffle, air turbulence will pick up condensed water and carry it downstream. If regular draining is a problem, install an automatic drain, or replace the assembly with a type that has an automatic drain feature.

A filter maintenance program should include physical inspection of the element. A clogged element may be subjected to pressure differential forces that exceed the element's structural limitations, causing the element to distort, collapse, or even break apart. Contaminants then bypass the element; in some cases, broken element pieces migrate downstream. An element that collapses but maintains its seal can cause a severe pressure loss in the downstream system.

It is impossible to accurately predict the service life of an air-filter element because of varying conditions of the incoming air and diverse operating cycles. However, most manufacturers mention weeks or even months of operation before cleaning or replacement should be necessary. Even if more frequent service should be required, the user still gets the degree of filtration specified, and the equipment is protected.

Cleanable elements are blown clean with compressed air or washed in water or solvent according to the manufacturer's directions. Good safety practice calls for the user to replace the dirty element with a spare, then clean the dirty element elsewhere, preferably in an enclosed area designed for that purpose.

Most filter designs make it possible to remove and replace a filter element without tools. Be sure to check element orientation and end-cap seating before reassembling. While removed, the bowl can be cleaned for better visibility.

Oil-absorbing elements act as a wick when wet, and feed oil back into the airstream. They should be changed before this condition occurs. Oil-absorbing elements function properly while wet, and need to be changed only when solid-particle accumulation generates an excessive pressure drop.

Aftercoolers and Air Dryers

Aftercoolers

As its name implies, an aftercooler cools compressed air after compression has been completed. This is accomplished by passing cooling water or air over the aftercooler chamber. Besides being the point where air cools, an aftercooler is also the place where some dirt and oil vapor fall out of suspension, and a good portion of entrained water vapor coalesces out. The aftercooler must have a moisture separator, preferably having an automatic drain.

We know that when a gas cools, its specific volume decreases. This change usually results in a change (decrease) in the pressure of the gas. Also, when air cools, its ability to carry water vapor decreases. In an aftercooler, compressed air and the water vapor it carries are cooled and its water content is made to condense into "rain." As air passes to the air receiver, it has much less potential energy than when it entered; it is cooler, cleaner, and holds less water. The air leaving the compressor is typically very humid. This high humidity requires an aftercooler to remove some of the water vapor.

In a typical water-type aftercooler, the direction of water flow is opposite that of the airflow. A good aftercooler will cool the air flowing through it to within 15°F of the cooling water temperature. It will also condense up to 90% of the water vapor originally contained in the air as it enters the receiver tank.

Water removal is as vital as heat removal; 100 ft³ of air after compression can release as much as 1.4 qt of water. Thus a modest-sized 100-scfm system could produce more than 50 gal of condensed water in a single 24-hr day. If demands of a system require drier air, the compressed air can then be put through the following processes:

1. Overcompression
2. Refrigeration
3. Absorption
4. Adsorption
5. Combination of the methods above

Overcompression

In the overcompression process, the air is compressed so that the partial pressure of the water vapor exceeds the saturation pressure. Then the air is allowed to expand, thereby becoming drier. This is the simplest method, but the power consumption is high. It is usually used for very small systems and therefore is not as common in industry as are the other methods.

Refrigeration

As mentioned, when the temperature of air is lowered, its ability to hold gaseous water is reduced. This is what takes place in an after-cooler. However, the typical minimum air temperature attainable in an aftercooler is limited by the temperature of the cooling water or air. If extremely dry air is needed, a refrigerant-type cooler is employed. In these devices, hot incoming air is allowed to exchange heat with the cold outgoing air in a heat exchanger. The lowest temperature to which the air is cooled is 32.4°F to prevent frost from forming. This type of air-drying equipment has relatively low initial and operating costs.

Absorption

One way water vapor in compressed air can be removed is by absorption. There are typically two basic absorption methods. In the first, water vapor is absorbed in a solid block of chemicals without liquefying the solid. The chemicals used in the solid insoluble type are typically dehydrated chalk and magnesium perchlorate. Another type uses deliquescent drying agents such as lithium chloride and calcium chloride, which react chemically with water vapor and liquefy as the absorption proceeds. These agents must be replenished periodically.

Some problems tend to exist with the deliquescent drying process because most of these drying agents are highly corrosive. Also, the desiccant pellets can soften and bake at temperatures exceeding 90°F. This may cause an increased pressure drop. In addition, a fine corrosive mist may be carried downstream with the air and corrode system components. However, this type of air dryer has the lowest initial and operating cost of the most common types. Maintenance is simple, requiring periodic replacement of the deliquescent drying agent.

Adsorption

Adsorption (desiccant) drying is another industrial method for drying air. Adsorption chemicals hold water vapor in small pores in the desiccant chemicals. Processes of this sort typically seek to make use of chemical such as silica gel (SiO_2) or activated alumina (Al_2O_3). This type of air drying is the most costly of the drying methods discussed here. This is because of moderate-to-high initial and high operating costs. However, maintenance costs may be lower than the absorption type, because there are no moving parts. Also, the replacement of the desiccant is eliminated.

Types of Dryers

In the broadest terms, there are three basic types of air dryers: deliquescent, regenerative desiccant, and refrigeration. *Deliquescent* dryers contain a chemical desiccant that reacts with the moisture in the air and absorbs it, regardless of whether it is already condensed or is still water vapor. The chemical is consumed in the drying (water-removing) process, and must be replenished periodically. The drain solution from these dryers contains both condensed water and some of the chemical. The dryer must be drained daily. Deliquescent dryers reduce the dew point of the air 15 to 25°F below the inlet air temperature, so if the incoming air has a temperature of 90°F, it will leave a deliquescent dryer at a dew point of about 65°F. Depending on operating conditions, some deliquescent dryers can produce dew points as low as 40°F.

Regenerative desiccant dryers remove water from air by absorbing it on the surface of a solid desiccant, usually silica gel, activated alumina, or molecular sieve. The desiccant does not react chemically with the water, so it need not be replenished, but it must be dried, or regenerated, periodically (see Figures 5.12 and 5.13).

Heatless regenerative dryers use two identical chambers filled with desiccant. As air moves up through one chamber and is dried, a portion of the dry, discharged air is diverted through the second chamber, reactivating the desiccant. The moisture-laden purge air is discharged to atmosphere. A short time later, air flow through the chambers is reversed.

Heated regenerative dryers use two identical chambers. In this type, however, air flows through one chamber until the desiccant has absorbed all the moisture it can hold, at which time airflow is diverted

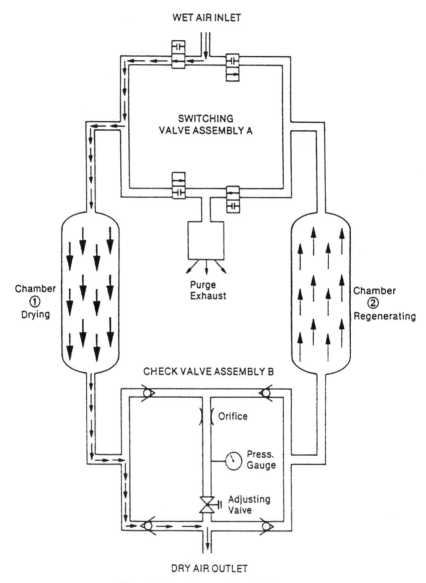

WET AIR INLET

SWITCHING
VALVE ASSEMBLY A

Chamber
①
Drying

Purge
Exhaust

Chamber
②
Regenerating

CHECK VALVE ASSEMBLY B

Orifice

Press.
Gauge

Adjusting
Valve

DRY AIR OUTLET

Figure 5.12. Regenerative air dryer.

to the second chamber. Heated outside air or an external source of heat
(steam or electric) then dries the desiccant in the first chamber. Because
desiccants have lower adsorption capacity at higher temperatures, the
desiccant bed must be cooled from the temperature it reached during

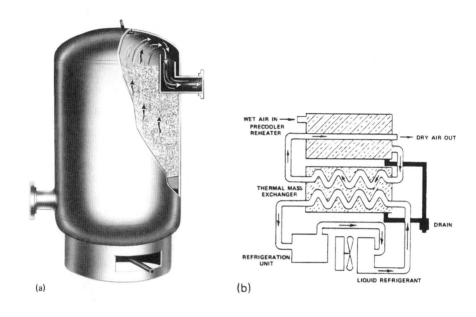

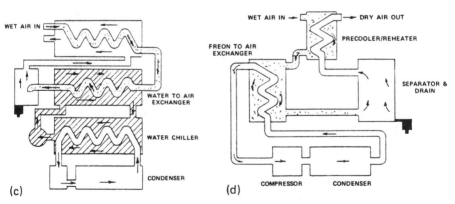

Figure 5.13.   Types of air dryers.

regeneration. The regeneration cycle in these dryers usually lasts several hours, divided between heating (75%) and cooling (25%) times.

Refrigeration dryers condense moisture from compressed air by cooling the air in heat exchangers chilled by refrigerants such as Freon gas. These dryers will produce dew points of 33 to 50°F at system operating pressure. Many refrigeration dryers reheat the compressed air after it has been dried, either with a heating element or by passing the cooled air back through the heat exchangers in contact with the hot incoming air. Reheating prevents condensation on air lines downstream from the dryer, and also helps cool incoming air. Refrigeration dryers must not be used where the ambient temperature is less than 35°F because lower temperatures will freeze the condensate, blocking air passages, and possibly damaging the evaporator.

Refrigeration dryers can be further classified into three basic types. *Tube-to-tube* refrigeration dryers operate by cooling a mass of aluminum granules or bronze ribbon that in turn cools the compressed air. These dryers can produce dew points of 35 to 50°F. *Water-chiller* refrigeration dryers use a mass of water for cooling. An extra heat exchanger is necessary to maintain chilled water flow through the condenser, as is a water pump. The dew point is 40 to 50°F. *Direct-expansion* refrigeration dryers use a Freon-to-air cooling process and achieve a dew point of 35°F under maximum operating conditions. No recovery period is necessary.

The tube-to-tube refrigeration dryer is a cycling dryer. A thermometer in the mass senses its temperature. As the temperature rises, a switch turns on the refrigeration unit; when the temperature drops to a cutoff point, refrigeration stops. Dryers of this type provide a dew point that varies within a few degrees. Water-chiller and direct-expansion refrigeration dryers run continuously. This continuous operation may add slightly to power requirements, and may result in freezing during light-load conditions if adequate controls for bypassing hot gas and/or reducing compressor capacity are not included. Table 5.2 compares the various types of compressed-air dryers.

How dry must the air be? The most important criterion in choosing an air dryer is the dew point, the temperature to which air can be cooled before water begins to condense from it. The required dew point of the air system determines how dry the air must be, and, to a great extent, which type of dryer you choose. Keep the following points in mind:

1. Dew point varies with pressure. For example, an atmospheric dew point of −12°F is equivalent to a pressure dew point of 35°F at 100 psig. Be sure that you know whether a manufacturer is specifying the dew point the dryer can attain at atmospheric pressure or at a typical system pressure such as 100 psig. You

Table 5.2
A Comparison of Air Dryers*

Type of dryer	Initial cost 1,000-scfm unit	Pressure dew point (100 psig) achieved w/100°F inlet air (°F)	Operational cost[a] per 1,000,000 ft^3 (including depreciation) at maximum flow: 100°F 100 psig inlet	Yearly costs per 100-hp compressor capacity[b]
Refrigeration	$6,000	40	$3.91	$185
Twin-tower desiccant				
Heated	8,000	0	6.64[c]	315
Heatless	7,000	−70	23.00[c]	1,091
Deliquescent desiccant				
Salt or urea	3,000	80–85	20.00[d]	949

[a] *Energy costs based on $0.035/kWh.*
[b] *Operating 80% of time, 8 hr/day, 260 days/year*
[c] *Includes cost of regenerating heat or air plus maintenance of prefilters/after-filters.*
[d] *Includes cost of replacement desiccant, freight, handling, storage, downtime, and maintenance of prefilters/afterfilters.*
* *Courtesy of Van Air Corp.*

can then determine what the minimum dew point will be at the system's operating pressure.

2. Required dew point varies with the application. If you are concerned primarily with preventing condensation in compressed-air lines, the lowest ambient temperature to which the air lines will be subjected will be the controlling factor (there may be fluctuations of ±35°F or more from summer to winter). However, for some applications, dew-point requirements will be more severe, possibly as low as −40°F at atmospheric pressure. Do not inject too great a safety factor by starting with a dew-point level that is not really needed. A safety margin of 20°F is about the maximum recommended.

3. You may require extremely low dew points at only a few isolated points. Consider using individual dryers at each point of use to attain these low dew points, in tandem with a less expensive dryer that will dry the air to less stringent requirements for use in the rest of the air system.

What flow capacity is needed? An air dryer must not only be able to dry the compressed air to the required dew point, it must also be able to handle the airflow required without causing excessive pressure drop. The flow capacity of a dryer depends on the operating pressure, inlet air temperature, ambient air or cooling water temperature, and required dew point. When any of these conditions change, the flow capacity of the dryer also changes. Most manufacturers can supply performance curves that show the relationship of their dryers' flow capacities to these four factors. They merit careful consideration.

Installation and Maintenance of Driers

Where you install an air dryer can affect how well it performs. The amount of maintenance a dryer requires can add to its total yearly cost in terms of both labor and materials. If all the compressed air will be used inside a building where temperature is maintained at a stable level, the required dew point can be fixed within a range of a few degrees. If, however, some or all of the compressed air is subjected to outdoor temperature variations, the required dew point can change from day to day, or even from hour to hour. There is an upper limit to ambient air temperature for refrigeration air dryers of about 100 to 110°F. Above this level, there is no efficient heat sink, and the dryer will not operate properly. Water-cooled condensers can tolerate higher ambients. Refrigerant air dryers should not be exposed to ambient temperatures much below 40 to 55°F without the addition of low-ambient-temperature controls.

When using deliquescent dryers, if the dryer is used in a central system, add bypass piping around the dryer to maintain the air supply while adding desiccant to the dryer. Also make certain there is no set of conditions (such as a valve that can be opened) that will reduce system pressure and cause high airflow velocity through the dryer, possibly carrying the chemical into the air lines (Figure 5.14).

Refrigeration and deliquescent dryers should be drained regularly, depending on the amount of moisture accumulation. Most refrigeration

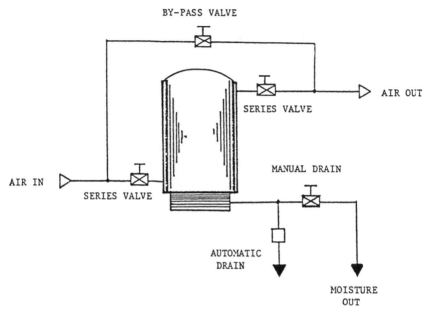

Figure 5.14. Dryer installation.

dryers have automatic drains. These are available as options on deliquescent dryers. Some refrigeration dryers require a prefilter to remove oil and dirt that can coat the inside of the dryer, lowering heat transfer. With regenerative desiccant dryers, oil from the compressor can coat the desiccant, rendering it useless; install equipment designed for oil removal ahead of the dryer.

Troubleshooting the Compressed-Air Dryer

Compressed-air-dryer problems generally fall into one of four types:

1. Water carryover
2. Oil carryover
3. Excessive desiccant usage
4. Excessive pressure drop

Often, the trouble encountered is the result of more than one of the problems. Thus the entire list should be considered.

Water Carryover Into the Downstream Air Line

Dryer undersize for the application. Perhaps the original estimate of airflow in the circuit was in error, or perhaps additional equipment has been added since the dryer was installed. This results in an undersized unit. The existing dryer need not be discarded. Drying capacity can be increased by adding a second dryer in parallel with the first. The new dryer may be the same or a different size.

High air temperature. The inlet air may be too warm for proper drying action. This may occur only at certain times of the day. A 24-hour check should be made by recording temperature every hour or even more often. First, feel the inlet pipe. If it is above body temperature by even a slight amount, this is a fairly sure indication of potential trouble, such as excessive desiccant use and water carryover.

Radiant heat. There may be radiant heat sources, such as hot air ducts, steam or hot water pipes, oven, eupola, or furnaces, near the upstream piping or near the dryer shell. Radiant heat picked up by the airstream causes caking of the desiccant near the bottom of the bed, even though the top of the bed looks normal. Radiant heat picked up by the shell may cause the desiccant to pull away from the sides of the vessel.

Oil Carryover Into the Downstream Air Line

Oil source. In nearly every case, unwanted oil in the air line comes from the air compressor. Desiccant dryers will usually precipitate only a very small amount of oil—that which will emulsify with the water that is removed. Any excess above this amount will collect on the surface of the pellets and can easily be observed by inspection of the bed through the access cover.

A compressor that is badly in need of an overhaul (pumping oil) will not provide satisfactory results with an air dryer. If the compressor uses the splash system of lubrication, and the oil carryover is excessive, the remedy is overhaul or replacement of the compressor. Oil-extracting filters installed ahead of the dryer may help to reduce the severity of the problem. Rotary compressors that inject oil as a lubricant into the compressor inlet are generally not satisfactory for use with air dryers. Oilless rotary compressors give very satisfactory results.

Excessive desiccant consumption. The specific amount of desiccant that should be used per day or per month cannot be firmly established in rule form. However, a good procedure is to keep a record of the desiccant quantity added over a period of time, and from this a weekly rate or monthly rate of consumption can be established. This should be done when the dryer is installed. Subsequently, if consumption increases, the various factors that cause excessive consumption can be investigated. These include drain stoppage, high inlet-air temperature, and intermittent or continuous overloading with excessive airflow.

Excessive pressure drop. If the dryer has been sized correctly, the pressure drop through it should generally not exceed 2 psi. Excessive pressure drop is nearly always caused by overloading of the dryer. The increased velocity of the air causes increased pressure drop. The overloading may occur either because of excessive air consumption downstream, or from low inlet pressure to the dryer.

Fluid Conductors and Connectors

Pipe, Tubing, and Hose

Many times the method of marrying interacting components of a system is a minor consideration. We know that there must be some sort of fluid-carrying link between components, but the way it is done is sometimes considered arbitrary. This attitude may result in a system that is inefficient, unsafe, unattractive, and difficult to service. Conductors of a fluid-power system are basically of three types: pipe, tubing, and hose.

The choice between pipe or tubing depends on system pressure and flow. The advantages of tubing include easier bending, fewer fittings, better appearance, better reusability, and less leakage. However, pipe is cheaper and will handle larger volumes under higher pressure. Pipe is also used where straight-line hookups are needed, and for more permanent installations.

In either case, the hydraulic lines must be compatible with the entire system. Pressure loss in the lines must be kept to a minimum for an efficient circuit.

Hose is flexible and its selection is generally based on the need for flexibility (swiveling, bending, pivoting, etc.)

Pipe

Pipe is a rigid conductor that is not intended to be bent or shaped into a desired configuration.

Pipe can be manufactured and purchased in a variety of materials, such as cast iron, steel, copper, aluminum, brass, and stainless steel. Pneumatic systems generally require a corrosion-resistant pipe. Hydraulic systems use steel pipe. Galvanized pipe is not recommended for use in hydraulic systems because the zinc coating of the pipe interacts unfavorably with oil or it may flake or scale and damage the valves and pump.

The inside diameter (ID) of a pipe, or any fluid-carrying conductor, is an important consideration. For if the ID is too small, a large amount of friction results, which translates into system inefficiency and wasted energy.

Wall thickness of a fluid conductor determines its pressure rating. Wall thickness of a pipe is identified by its schedule number. There are ten schedule numbers ranging from 10 to 160. Schedule numbers 40, 80, and 160 are the most commonly used pipe in a fluid-power system. Schedule 40 pipe has a wall thickness rated for low pressure, schedule 80 is for high pressure, and schedule 160 is for very high pressure (Table 6.1).

Table 6.1
Dimensions of Welded and Seamless-Steel Pipe

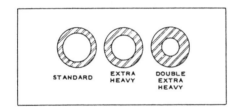

STANDARD EXTRA HEAVY DOUBLE EXTRA HEAVY

Listed as Std., X or XX   Listed by schedule numbers

Nominal Size	Pipe O.D.	Standard	Extra Heavy	Double Extra Heavy	Sched. .20	Sched. 30	Sched. 40	Sched. 60	Sched. 80	Sched. 100	Sched. 120	Sched. 140	Sched. 160
1/8	.405	.269	.215				.269		.215				
1/4	.540	.364	.302				.364		.302				
3/8	.675	.493	.423				.493		.423				
1/2	.840	.622	.546	.252			.622		.546				.466
3/4	1.050	.824	.742	.434			.824		.742				.614
1	1.315	1.049	.957	.599			1.049		.957				.815
1-1/4	1.660	1.380	1.278	.896			1.380		1.278				1.160
1-1/2	1.900	1.610	1.500	1.100			1.610		1.500				1.338
2	2.375	2.067	1.939	1.503			2.067		1.939				1.689
2-1/2	2.875	2.469	2.323	1.771			2.469		2.323				2.125
3	3.500	3.068	2.900				3.068		2.900				2.624
3-1/2	4.000	3.548	3.364				3.548		3.364				
4	4.500	4.026	3.826				4.026		3.826		3.624		3.438
5	5.563	5.047	4.813	4.063			5.047		4.813		4.563		4.313
6	6.625	6.065	5.761				6.065		5.761		5.501		5.189
8	8.625	8.071	7.625		8.125	8.071	7.981	7.813	7.625	7.439	7.189	7.001	6.813
10	10.750	10.192	9.750		10.250	10.136	10.020		9.750	9.564	9.314	9.064	8.500
12	12.750	12.080	11.750		12.250	12.090	11.934	11.626	11.376	11.064	10.750	10.500	10.126

Pipe connections are made by means of threaded joints. To join a pipe to a component or pipe fitting, a threaded end of a pipe (male end) is screwed into a female thread of a component or fitting.

Pipe threads have another function besides joining, and that is sealing. To form a seal, pipe threads are tapered on the diameter $1/16$ in. per inch of length. As a male pipe thread tightens into a female pipe thread, the metal-to-metal interference that occurs is intended to form a seal. In actual practice, however, temperature changes, vibration, system shock, or an imperfect thread may tend to destroy the sealing capabilities of the metal-to-metal joint. For this reason, thread sealant of various types are commonly used to help make and maintain the seal.

Two pipe-thread configurations are generally used in fluid systems—standard pipe thread and dry-seal pipe thread. The difference between the two configurations is shown on Figure 6.1. As can be seen, the dry-seal thread eliminates the spiral clearance left after tightening a standard pipe thread connection. This reduces the need for a sealant to make the joint leak-tight (such as pipe dope, teflon tape, etc.). These sealants, when carelessly applied, can enter the system and harm the sensitive components.

Tubing

Tubing is a semirigid fluid conductor, which is customarily bent into a desired shape. The use of tubing gives a system with a neat appearance; a system less susceptible to leaks and vibration; and a system whose conductors can be easily removed and replaced for maintenance purposes.

Tubing is made from a variety of materials including steel, copper, brass, aluminum, stainless steel, and plastic. Hydraulic systems generally use steel tubing. Just as with other tubular materials, tubing is measured by its outside diameter, inside diameter, and wall thickness. The inside diameter determines how much fluid flow the tubing can efficiently pass. Wall thickness determines the maximum pressure at which the material can be used for any given inside diameter.

Copper. The use of copper is limited to low-pressure hydraulic systems where vibration is limited. Also, copper tends to become brittle when flared and subjected to high heat.

Aluminum. This tubing is also limited to low-pressure use, yet has good flaring and bend characteristics.

Plastic. Plastic tubing lines are made from a variety of materials; nylon is the most suitable. For use in low-pressure hydraulic applications only.

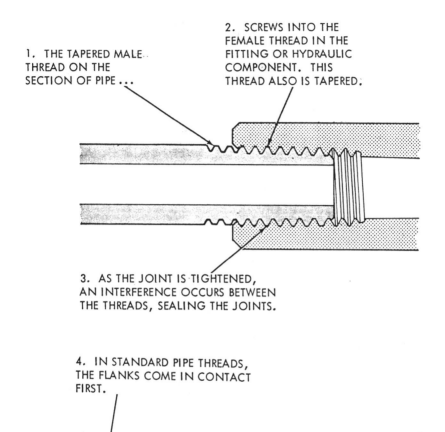

1. THE TAPERED MALE
THREAD ON THE
SECTION OF PIPE . . .

2. SCREWS INTO THE
FEMALE THREAD IN THE
FITTING OR HYDRAULIC
COMPONENT. THIS
THREAD ALSO IS TAPERED.

3. AS THE JOINT IS TIGHTENED,
AN INTERFERENCE OCCURS BETWEEN
THE THREADS, SEALING THE JOINTS.

4. IN STANDARD PIPE THREADS,
THE FLANKS COME IN CONTACT
FIRST.

5. THERE CAN BE A
SPIRAL CLEARANCE AROUND
THE THREADS.

6. IN DRY-SEAL THREADS,
THE ROOTS AND CRESTS
ENGAGE FIRST, ELIMINATING
SPIRAL CLEARANCE.

Figure 6.1. Pipe-thread configurations.

Steel. Tubing constructed of cold-drawn steel has become the accepted standard in hydraulics where high pressures are encountered. There are two types of steel tubing: seamless and electric welded. Seamless is produced by a cold drawing of pierced or hot extruded billets. Welded is made by forming a cold-rolled piece of steel into a tube, then welding and drawing it.

Tubing size is indicated by its actual outside diameter. For example, $\frac{1}{8}$ in. tubing has an actual outside diameter of $\frac{1}{8}$ in. Tubing inside depends on wall thickness. The recommended wall thickness of tubing made of the commonly used materials and the working pressure are given in Table 6.2.

Tubing is connected to system components and to other conductors by means of tube fittings. Basically, there are two types of tube fittings: flared and flareless.

Hose

Hose is a flexible fluid conductor that can adapt to machine memebers that move. Hoses consist of three basic parts (Figure 6.2). The inner tube, which carries the fluid; the reinforcement, which provides the resistance to pressure; and the cover, which protects the hose from outside elements. The inner tube must be flexible and compatible with the fluid being carried. Synthetic rubber, thermoplastic, nylon, and teflon are some of the compounds used. The reinforcement layer provides the strength to withstand the system's pressure. One or more layers of cotton, synthetic yarn, or wire are braided or spirally wound or wrapped over the inner tube (Figure 6.3). The spiral wire reinforcement provides better flexibility, more resistance to failure, and a greater maximum working pressure than does braid reinforcement. It is also more expensive to manufacture. The hose cover protects the reinforcement from abrasion and corrosion. The cover carries a name, part number, the hose size, SAE number or rating, and date of manufacture. This information on the cover is referred to as the "layline" in the industry.

The Society of Automotive Engineers (SAE) has recommended standards for the hydraulic-hose industry. 100 "R" numbers from R1 to R11 are used to indicate hose performance capabilities and construction. For example, 100R1 is rubber hose with one layer of wire-braid reinforcement. It is suitalbe for most medium-pressure applications. 100R11 has a heavy wire-wrap reinforcement and is used in extremely high-pressure systems. Some specification examples for the SAE standard construction details are contained in Figure 6.4.

Hose size is customarily given by a dash number, which offers some identification as to its inside diameter. Dash numbers indicate 16ths of an inch. A −8 is equivalent to $\frac{8}{16}$ in. or $\frac{1}{2}$ in. A −8 hose means that the inside hose diameter is $\frac{1}{2}$ in. or a little less.

(text continued on page 232)

Table 6.2
Recommended Wall Thickness of Tubing

TUBING		1010 and 1015 STEEL, HALF HARD COPPER TUBING and BUNDYWELD													
		4/1 SAFETY FACTOR				5/1 SAFETY FACTOR			7.5/1 SAFETY FACTOR				10/1 SAFETY FACTOR		
		Working Pressure in psi.				Working Pressure in psi.			Working Pressure in psi.				Working Pressure in psi.		
Tube O.D.	Fitting Size	1000	2000	3000	5000	1000	2000	3000	1000	2000	3000	5000	1000	2000	3000
1/8	2	.020	.020	.020	.025	.020	.020	.020	.020	.020	.025	.042	.020	.025	.035
3/16	3	.020	.020	.020	.035	.020	.020	.028	.020	.025	.042	.065	.020	.035	.056
1/4	4	.020	.020	.028	.049	.020	.022	.035	.020	.035	.058	.095	.022	.049	.072
5/16	5	.020	.025	.035	.056	.020	.028	.042	.022	.042	.065	.109	.028	.056	.083
3/8	6	.020	.028	.042	.072	.020	.035	.049	.025	.058	.083	.134	.035	.072	.109
1/2	8	.020	.042	.056	.095	.025	.049	.065	.035	.072	.109	.220	.049	.095	.134
5/8	10	.025	.042	.072	.120	.032	.058	.083	.042	.095	.134	.220	.058	.120	.180
3/4	12	.028	.058	.083	.134	.035	.072	.109	.058	.109	.148	.259	.072	.134	.203
7/8	14	.035	.072	.095	.165	.042	.083	.120	.065	.120	.180	.320	.083	.165	.238
1	16	.042	.083	.109	.180	.049	.095	.134	.072	.134	.203	.350	.095	.180	.259
1-1/4	20	.049	.095	.134	.238	.058	.120	.165	.095	.180	.259	.450	.120	.238	.350
1-1/2	24	.058	.120	.165	.284	.072	.134	.203	.109	.203	.320	.500	.134	.284	.450
2	32	.072	.148	.220	.375	.095	.180	.259	.134	.284	.400	....	.180	.375	....

TUBING		ANNEALED COPPER		5250 ALUMINUM		3S ALUMINUM TUBING			4130 STEEL	
		5/1 SAFETY FACTOR		5/1 SAFETY FACTOR		5/1 SAFETY FACTOR			5/1 SAFETY FACTOR	
		Working Pressure in psi.		Working Pressure in psi.		Working Pressure in psi.			Working Pressure in psi.	
Tube O.D.	Fitting Size	500	1000	500	1000	250	500	1000	3000	5000
1/8	2	.020	.020	.018	.018	.018	.018	.020	.018	.018
3/16	3	.020	.020	.018	.018	.018	.018	.030	.018	.018
1/4	4	.020	.020	.018	.025	.018	.018	.042	.018	.020
5/16	5	.020	.028	.018	.032	.018	.025	.049	.018	.028
3/8	6	.020	.032	.018	.035	.018	.028	.065	.018	.032
1/2	8	.022	.042	.025	.049	.020	.042	.083	.025	.042
5/8	10	.028	.056	.028	.065	.025	.049	.109	.032	.049
3/4	12	.....	.....	.....	.....	....	.....	.....	.035	.065
7/8	14	.....	.....	.....	.....	....	.....	.....	.042	.072
1	16	.....	.....	.....	.....	....	.....	.....	.049	.083
1-1/4	20	.....	.....	.....	.....	....	.....	.....	.065	.109
1-1/2	24	.....	.....	.....	.....	....	.....	.....	.072	.120
2	32	.....	.....	.....	.....	....	.....	.....	.095	.159

TUBING		½ STAINLESS STEEL				ANNEALED STAINLESS STEEL			
		5/1 SAFETY FACTOR				5/1 SAFETY FACTOR			
Tube O.D.	Fitting Size	1000	2000	3000	5000	1000	2000	3000	5000
1/8	2	....	.018	.018	.018	.020	.020	.020	.020
3/16	3	....	.018	.018	.025	.020	.020	.020	.028
1/4	4	....	.018	.018	.032	.020	.020	.022	.042
5/16	5	....	.018	.025	.042	.020	.020	.025	.049
3/8	6	....	.018	.028	.049	.020	.022	.032	.065
1/2	8	....	.025	.042	.065	.020	.025	.042	.083
5/8	10	.018	.032	.049	.083	.020	.032	.058	.109
3/4	12	.020	.042	.058	.095	.025	.042	.065	.120
7/8	14	.022	.049	.065	.109	.028	.049	.072	.148
1	16	.025	.056	.083	.134	.032	.058	.083	.165
1-1/4	20	.032	.065	.109	.165	.042	.072	.109	.203
1-1/2	24	.042	.083	.120	.203	.049	.095	.134	.238
2	32	.049	.109	.165	.250	.065	.120	.180	.320

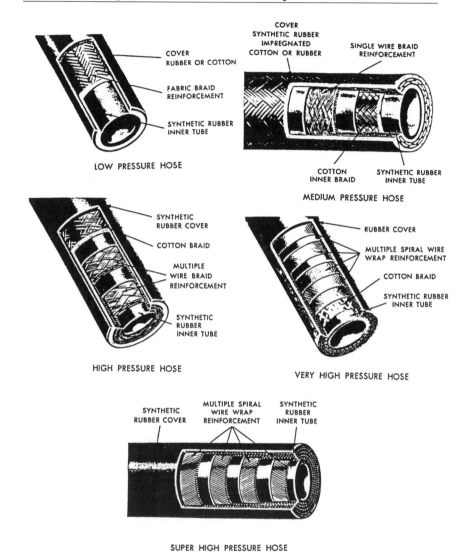

Figure 6.2. Hose construction.

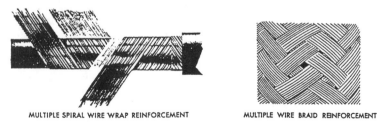

Figure 6.3. Braided and spiral hose construction.

SAE 100R1

Type A — This hose shall consist of an inner tube of oil resistant synthetic rubber, a single wire braid reinforcement, and an oil and weather resistant synthetic rubber cover. A ply or braid of suitable material may be used over the inner tube and/or over the wire reinforcement to anchor the synthetic rubber to the wire.

Type AT — This hose shall be of the same construction as Type A, except having a cover designed to assemble with fittings which do not require removal of the cover or a portion thereof.

SAE 100R2

The hose shall consist of an inner tube of oil resistant synthetic rubber, steel wire reinforcement according to hose type as detailed below, and an oil and weather resistant synthetic rubber cover. A ply or braid of suitable material may be used over the inner tube and/or over the wire reinforcement to anchor the synthetic rubber to the wire.

Type A — This hose shall have two braids of wire reinforcement.

Type B — This hose shall have two spiral plies and one braid of wire reinforcement.

Type AT — This hose shall be of the same construction as Type A, except having a cover designed to assemble with fittings which do not require removal of the cover or a portion thereof.

Type BT — This hose shall be of the same construction as Type B except having a cover designed to assemble with fittings which do not require removal of the cover or a portion thereof.

SAE 100R3

The hose shall consist of an inner tube of oil resistant synthetic rubber, two braids of suitable textile yarn, and an oil and weather resistant synthetic rubber cover.

SAE 100R4

The hose shall consist of an inner tube of oil resistant synthetic rubber, a reinforcement consisting of a ply or plies of woven or braided textile fibers with a suitable spiral of body wire, and an oil and weather resistant synthetic rubber cover.

SAE 100R5

The hose shall consist of an inner tube of oil resistant synthetic rubber and two textile braids separated by a high tensile steel wire braid. All braids are to be impregnated with an oil and mildew resistant synthetic rubber compound.

SAE 100R6

The hose shall consist of an inner tube of oil resistant synthetic rubber, one braided ply of suitable textile yarn, and an oil and weather resistant synthetic rubber cover.

SAE 100R7

The hose shall consist of a thermoplastic inner tube resistant to hydraulic fluids with suitable synthetic fiber reinforcement and a hydraulic fluid and weather resistant thermoplastic cover.

SAE 100R8

The hose shall consist of a thermoplastic inner tube resistant to hydraulic fluids with suitable synthetic fiber reinforcement and a hydraulic fluid and weather resistant thermoplastic cover.

Figure 6.4.   Hose construction standards (SAE).

(text continued from page 228)

The type and number of reinforcing layers of a hose determine under what system conditions it may be used. Hose-pressure classifications are suction, medium pressure, high pressure, and very high pressure.

Hose is connected to system components and to pipe or tubing by means of hose fittings. Hose fittings are classified as permanent or reusable.

With a permanent hose fitting, the hose is inserted into the fitting between nipple and socket. The socket is then crimped or swaged to hold the hose. Barbs on the outside diameter of the nipple ensure that the hose is securely held in place.

Reusable hose fittings are screwed or clamped to a hose end. They can be removed from a worn hose and reassembled onto a replacement hose.

Connectors

Tube Fittings

Tubing is the logical choice for flow lines in many hydraulic and pneumatic systems. The reasons for this preference vary, but more often than not the determining factor is the ease of assembly and disassembly associated with tubing lines.

A key assembly element for tubing-connections is the tube fitting. These fittings are produced in many types, styles, forms, and materials, offering a wide range of choice in solving specific tubing-connection problems.

To determine which fitting is best for a specific job, several factors must be taken into account. Typical questions that should be answered are: What pressure and temperature conditions are involved? What is the nature of the working fluid? In what kind of environment will the system be used?

For example, consider a hydraulic system that is to be operated in a highly corrosive atmosphere. In a normal atmosphere, such a system would call for steel fittings and tubing; under the circumstances, a change in fitting and tubing material is dictated to one compatible with the corrosive atmosphere.

In certain types of hydraulic systems, the operation of various components can cause dangerous pressure surges or set up severe shock conditions. It is not uncommon for a hydraulic unit, operating at 1000 psi, to develop shock loads of 5,000 to 8,000 psi when solenoid valves are used. In a similar situation, a 5,000 psi system was actually found to peak out at 30,000 psi, causing failure of a backwelded pipe system by stress cracking of the pipe.

This type of failure can be both dangerous and costly. It could have been averted by a thorough examination of system-operating conditions. Many more examples of this type could be cited.

Two basic types of fittings are in common use:

1. Flared
2. Compression, or flareless

Some of these fittings have been standardized. Many have not. In certain instances, operating characteristics and assembly features overlap, offering optional choices in fitting selection.

Flared Fittings

Fittings of the flared type can be further classified into two categories: the 45° style, and the 37° style. The angle in each case refers to the slope of the flared-tube surface with the center line of the tubing. Both of these fitting styles have been standardized by the SAE.

The 37° flared fitting is a refinement of the 45° design. Two versions of this fitting are available. One is a two-piece unit. The other 37° fitting style is a three-piece arrangement, which is made up of an externally threaded body similar to that of the two-piece version; a shouldered sleeve insert with the 37° mating tapered seat on one end, and an internally threaded sleeve nut that is assembled on the body thread to lock the sleeve in proper seated position. When the sleeve is seated, the untapered end extends through the nut for added tube support (see Figure 6.5).

The three-piece design is a more rugged construction that is suited for high-pressure service and severe operating conditions, including vibration, mechanical strain, and elevated temperatures. These fittings produce a fluid-tight joint that will hold beyond the burst pressure of conventional tubing lines. This fitting style is used almost exclusively for hydraulic and pneumatic systems in aircraft.

The effectiveness and reliability of the fittings depends primarily upon the quality of the tubing flare. The 37° fittings are available in steel, stainless steel, and brass. Figure 6.6 depicts good and bad flared tubing ends, along with the reasons for the condition.

A. During assembly the nut drives the sleeve forward against the rear of the flared tube end.

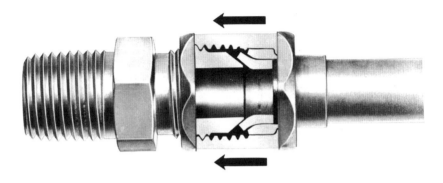

B. The sleeve and flared tube continue forward until the front of the flare is clamped in a compression seal against the mating 37° nose angle of the fitting body.

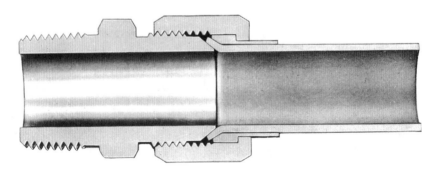

Figure 6.5. Flared-tubing fitting.

Compression Fittings

The need for a tube fitting that would eliminate the separate flaring operation was recognized some 70 years ago when the first compression (flareless) fitting (Figure 6.7) was introduced. Although

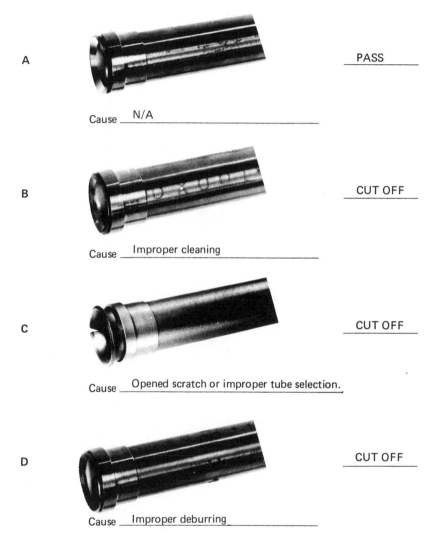

A   PASS

Cause   N/A

B   CUT OFF

Cause   Improper cleaning

C   CUT OFF

Cause   Opened scratch or improper tube selection.

D   CUT OFF

Cause   Improper deburring

Figure 6.6. Flared-tubing quality.

construction details vary considerably, operation of all of these fittings is based on the same general idea: as the nut is tightened on the body of the fitting, a compression grip is developed by a mechanical locking device to hold the end of the tubing in place.

From a production view, the compression fitting offers several attractive features. Tube preparation is kept to a minimum and usually

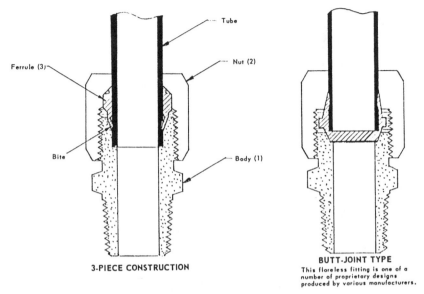

Figure 6.7. Compression-type fitting.

can be handled with conventional equipment. The need for special tube flaring equipment is eliminated. The assembly procedure for tubing connections becomes a relatively simple operation that can be handled by semiskilled operators, with little special training.

Fitting Forms

Each fitting type is usually avail ble in many different forms (Figure 6.8) to meet the various connection requirements encountered in tubing layouts. These forms include straights, elbows, tees, crosses, and other configurations for pipe-to-tube, tube-to-tube, tube-to-component, and port connections. In some instances, the range of available parts encompasses "jump sizes," such as ³/₈ in. to ½ in., and others.

Straight-Thread, O-Ring Type Fitting

It used to be that when leaks occurred in a hydrualic system, it was pretty much standard practice to grab a wrench and a can of pipe dope and set to work fixing the culprit. Unfortunately, these attempts to overcome leakage frequently compounded the problem. Pipe fittings were overtightened, resulting in damaged threads and in cracked or

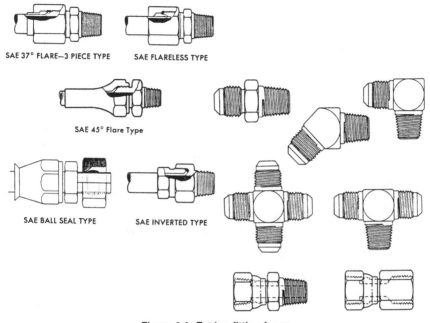

Figure 6.8. Tubing fitting forms.

distorted valve bodies or other components in the circuit. In addition, positioning requirements of tapered-pipe-thread elbows and tees often forced overtightening, with inevitable damage.

In an effort to solve these hydraulic-circuit leakage problems, the O-ring-type fittings were developed (Figure 6.9). These type fittings overcome the problems encountered with pipe fittings; there are at least five important reasons for this:

1. **No casting distortion.** SAE straight-thread connector fittings shoulder securely, providing a positive seal and a tight, mechanical joint without the hazard of leak-producing distortion.
2. **Exact positioning without "backing-off" or overtightening.** With straight-thread elbow or tee fittings, you can turn the fitting to the exact position required, then tighten the locknut down securely, assuring a leakproof seal.
3. **"Pipe dope" is eliminated.** Positive seal of straight-thread fittings in an SAE O-ring boss does away with the need for messy "pipe dope" (or other auxiliary sealing devices) and the danger that it may contaminate the hydraulic system.

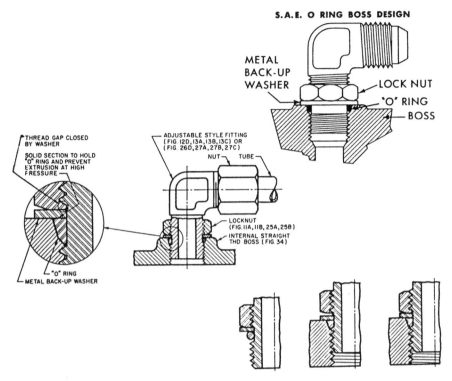

Figure 6.9.   SAE O-ring-style fitting.

4. **No temperature or shock leaks.** With straight threads you elimi-nate the danger of leaks due to temperature changes or high shock conditions.
5. **Mechanically rigid connection.** Proper torque can be applied because in straight-thread ports the fittings are working on full, perfect threads. This results in a solid connection that is completely rigid.

Hose Couplers and Fittings

Hose is connected to system components and to pipe or tubing by means of hose fittings. Hose fittings are classified as permanent or reusable (Figure 6.10).

With a permanent hose fitting, the hose is inserted into the fitting between nipple and socket. The socket is then crimped or swaged to hold the hose. Barbs on the outside diameter of the nipple ensure that the hose is securely held in place.

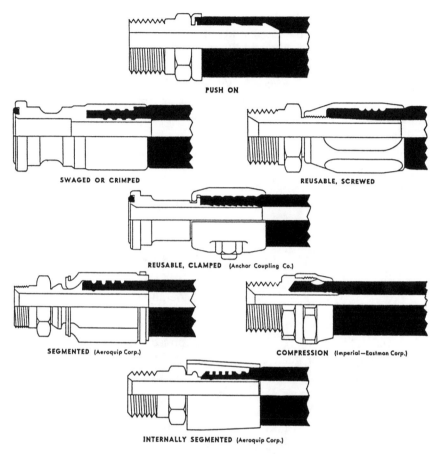

Figure 6.10.   Typical hose-fitting style.

Reusable hose fittings are screwed or clamped to a hose end. They can be removed from a worn hose and reassembled onto a replacement hose.

A skive-type fitting is a screw-on design for use on hoses with a thick outer cover. This cover is removed (skived) from the ends prior to fitting installation. A no-skive-type fitting is a screw-on design for use on hose with a thin outer cover. This cover does not require removal prior to fitting installation.

A clamp-type fitting is designed with a barbed nipple, which is inserted into a hose. Two clamp halves are then bolted together to provide a leakproof grip.

Hose couplers included two types (Figure 6.11): *Fittings* (part of the hose, they have a socket and nipple or sleeve) and *Adaptors* (a separate part for joining the hose fitting to another line). Fittings and adapters are called either male or female couplers. The hollow female coupler mates with the male type.

Hose couplers are made of steel, brass, stainless steel or, in a few applications, plastic. Steel is generally used because it withstands both high pressures and heat.

Hose fittings can seal in many ways. The five major methods are shown in Figure 6.12.

Besides straight fittings, elbow fittings are also available. Elbows should be used for access to hard-to-reach connections and for special routing problems.

Installation Considerations

Pipe Installation Considerations

Threading of pipe requires removal of metal. This means bare metal is exposed at the thread. When a threaded connection is made, it is advised that some sort of protective and sealant compound be applied to the threads to avoid corrosion and effect a seal. Protecting pipe threads from corrosion will help in the event the joint has to be disconnected.

When any type of sealant and protective coating is applied to a pipe thread, the coating should be applied up to the second pipe thread from the end only. This avoids contaminating the system with the coating material.

There is a caution when screwing in a pipe. Since pipe threads are tapered, the more a pipe is screwed into a component housing or fitting, the more likelihood of rupturing the housing or fitting from the wedging action of the joint.

The ends of sections of pipe may be threaded or welded to flanges. In either case, the ends must be cleaned when the operation is complete. Chips, burrs, and cutting compound must be removed from threads. Loose weld beads and spatter must be cleaned off before the pipe section is plumbed into the system.

Dryseal threads require no sealer, and are the preferred type for use in hydraulic systems. Tapered threads should not be assembled without a sealant. Teflon tape is now widely used for this purpose in hydraulic systems. It is neat, inexpensive, and easy to apply, causes

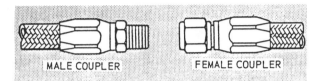

Figure 6.11.   Male and female hose couplers.

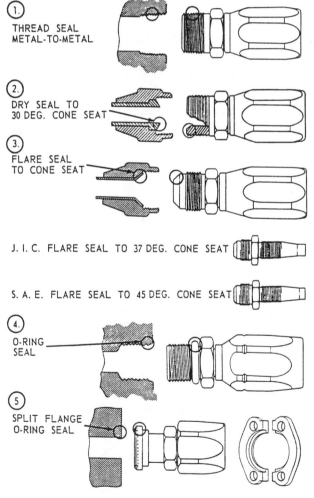

Figure 6.12.   Hose connection types.

no damage to connection, and will withstand temperatures up to 600°F. However, the worker using the tape must be careful to leave the first two threads of each joint bare, to ensure that fragments of tape will not enter the system.

Pipe should be supported with hangers and brackets that will help reduce vibration throughout the system; sometimes it is necessary to install small accumulators in the lines to act as shock suppressors, or to replace short lengths of pipe with hydraulic hose. If hose is used, its ID should be approximately that of the pipe it replaces.

Only clean pipe should be used in hydraulic systems, and even clean pipe should be flushed before installation. Rust and pipe scale can clog the orifices of control valves and score the walls of cylinders. Pipe salvaged from the scrap pile should never be employed.

When a hydraulic system is plumbed with rigid pipe, flow valves and pressure controls are often supported by the pipe. They should be provided with unions so that they can be removed for servicing without disturbing the surrounding piping.

Tubing Installation Considerations

The great advantage tubing has over pipe is that is can be bent. Each such bend eliminates a pipe elbow and two joints. Furthermore, the longer-radius bends provided by tubing result in less loss of energy. However, tubing must be properly bent, so that is does not wrinkle or flatten and restrict flow. Both power and manual benders can make good bends. Figure 6.13 shows good and bad bends, along with the proper bend radius for various tube sizes. The radius of the smallest bend, from the center to the center line of the tubing, should not be less than three times the outside diameter (OD) of the tubing.

Tubing can be cut with a power tube cutter, a manual tube cutter, or a hacksaw. All burrs must be removed after cutting. If a hacksaw has been used, the ends of the tubing should be filed square with the bore, and all filings removed from the interior of the tubing.

Although bending and cutting hydraulic tubing is fairly simple, designing a good installation requires some thought. Following are some pointers:

- Eliminate stress from the tubing lines. Support long runs with brackets or clips. If fluid shock is anticipated, use resilient material between the tubing and the bracket. Do not hang heavy controls from tubing (Figure 6.14).

STANDARD BEND RADII

Tube Size No.	Tube Outside Diameter	Standard Radius R
2	1/8	3/8
3	3/16	7/16
4	1/4	9/16
5	5/16	11/16
6	3/8	15/16
8	1/2	1 1/2
10	5/8	2
12	3/4	2 1/2
14	7/8	3
16	1	3 1/2

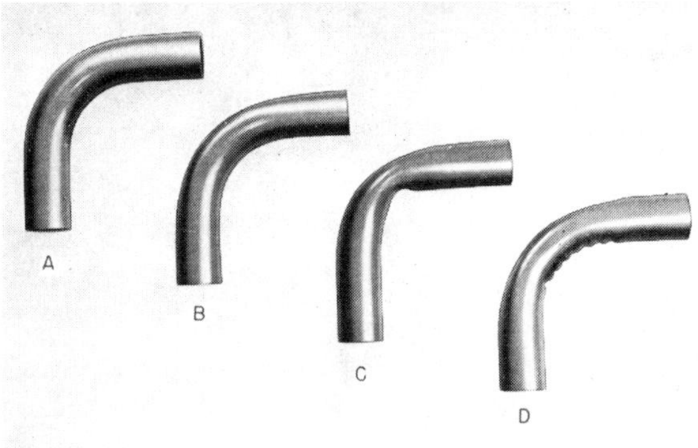

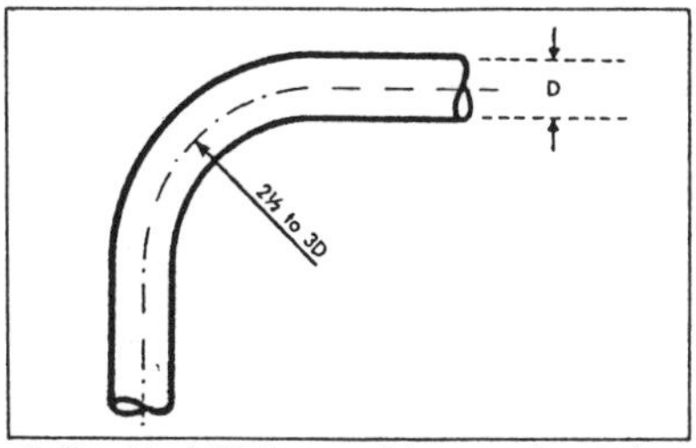

Radius of bend (R) is measured from centerline of tube.

Figure 6.13. Tube bends.

- Make certain the tubing is strong enough to withstand the system's pressure. Remember, some systems have high surge pressure that far exceeds the normal working pressure.
- Do not make straight-line connections on short assemblies. Even on longer lengths, straight-line connections are not recommended (Figure 6.15).

Figure 6.14. Pipe and tubing clamps. (Courtesy of Hycon Corporation.)

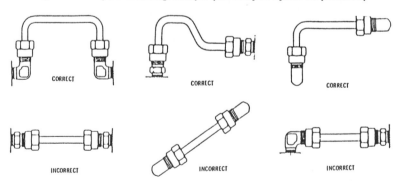

Figure 6.15. Tubing installations.

- Keep the system symmetrical if at all possible. Symmetrical tubing layouts present a workmanlike appearance.
- Allow at least twice the length of the nut from the end of the tube to the start of a bend to eliminate problems in assembly and disassembly. A longer straight section is even more desirable.

- Make assemblies so that they can be inserted in the fittings without tension or distortion.
- Do not hammer the tube assembly to get it into place. Hammering causes leaks and failures.
- Protect the tubing runs from forklift trucks, falling objects, and other dangers. Remember, unless a fire-resistant fluid is used in the system, a broken hydraulic line presents a real fire hazard.
- Do not run tubing across the floor where there is any possibility of traffic moving over it. If tubing *must* be placed on the floor surface, provide a protective cover.
- See that drain lines are lower than the controls to be drained. If possible, have all the lines between the reservoir and the machine sloped slightly toward the reservoir. Air trapped in the system can then work its way toward the high points in the piping, where it can be purged through air-relief valves.

When using flare-type fittings, it is important to remember the effects of the tubing wall thickness, as shown in Figure 6.16. Heavier wall tubing tends to reduce sealing surface and thread engagement length. The flared connection must be tightened properly. Overtightening will cause leakage, just as does undertightening. It is best to tighten the connection to a specific torque value specified by the manufactuer. However, the recommended turns of tightness, shown in Table 6.3, can be used, and these will produce safe, leak-free connections.

Hose Installation Considerations

During operation, hose should not be scraped or chafed. This eventually weakens the hose. Hose should not be twisted during installation or system operation. Twisting a hose reduces its life considerably, and may help in loosening hose fittings. A twisted hose can be easily detected by examining the line along the hose cover. It should not appear twisted.

To determine what hose and coupling you need, look in the appropriate parts or service manual for the equipment that is being repaired. Here you should be able to locate the part number and/or more importantly, the performance requirements, to help you make the proper selection.

If a hose fails simply because of old age, check the lay line for the SAE 100R rating number and hose size. Your job becomes easy here— just replace it with the same type of hose.

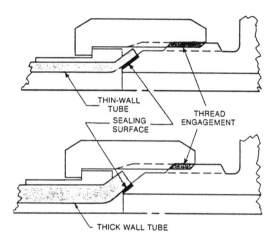

Figure 6.16. Effect of tubing wall thickness with flare fitting.

For best performance, a hose should be selected to meet the service conditions under which it is to be used. Before deciding on the size and type, you must consider the pressure, temperature, fluid type, and possible contaminants. Catalogs supply information on operating and burst pressure, temperature ranges, bend radius, and fluid compatibility, which can make selecting a replacemnet hose easier. If you're not sure which hose to pick, select a higher pressure hose than you feel the job needs. It's better to spend a few extra dollars than to risk an early failure.

Once you have selected the correct hose and couplings, they must be assembled properly. First, hoses must be cut to the proper length. The hose length equals the overall length of the assembly minus the distance the couplings extend beyond the ends of the hose. The cut must be clean and square in order to seat properly. Be careful of the hose ends; those exposed wires are sharp. Under pressure, hoses can shorten up to 4% of their total length or lengthen by as much as 2%. A 50-in. assembly, for example, can shorten as much as 2 in. Thus, provide enough slack to account for this change in dimension.

Hoses that will not be used immediately should have their ends closed with plastic plugs or suitably threaded caps. They should be stored in a cool, dark, dry area away from electrical equipment. Short assemblies should be laid straight. Longer assemblies can be coiled, provided the diameter of the bend is 3 ft or more. The hose covers should be protected.

Table 6.3
Tightening Flare-Type Fittings

Line Size (Outside Diameter)	Flare Nut Size (Across Flats)	Tightness (Ft-lbs)	Recommended Turns of Tightness (After Finger Tightening)	
			Original Assembly	Reassembly
$3/_{16}$"	$7/_{16}$"	10	$1/_3$ Turn	$1/_6$ Turn
$1/_4$"	$9/_{16}$"	10	$1/_4$ Turn	$1/_{12}$ Turn
$5/_{16}$"	$5/_8$"	10–15	$1/_4$ Turn	$1/_6$ Turn
$3/_8$"	$11/_{16}$"	20	$1/_4$ Turn	$1/_6$ Turn
$1/_2$"	$7/_8$"	30–40	$1/_6$ to $1/_4$ Turn	$1/_{12}$ Turn
$5/_8$"	1"	80–110	$1/_4$ Turn	$1/_6$ Turn
$3/_4$"	$11/_4$"	100–120	$1/_4$ Turn	$1/_6$ Turn

Plant engineers should keep the following pointers in mind when planning hose installations:

- Choose an assembly that will cause little pressure loss.
- Use only hoses intended for suction service on the suction side of the pump. Cavitation will result if the hose collapses.
- Allow plenty of hose in applications subject to substantial amounts of flexing. Often, 90° adapters can be used to reduce the flexing required of the hose.
- Consider the ambient temperatures to which the hose will be subjected. Special hoses for high and low temperatures are available. In some applications, fire sleeves will be needed to protect the hose cover.
- Make certain the hose cover will withstand chemicals to which it is likely to be exposed.

Follow these six basic rules when installing a hose (Figure 6.17):

1. **Avoid taut hose.** Even where the hose ends do not move in relation to each other, allow some slack to prevent strain. Taut hoses tend to bulge and weaken under pressure.
2. **Avoid loops.** Use angled fittings to prevent long loops. Doing this cuts down the length of hose needed and makes a neater installation.

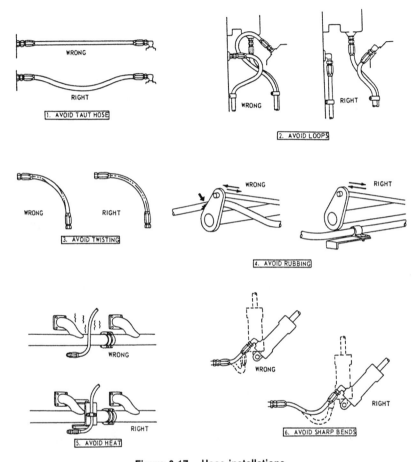

Figure 6.17.   Hose installations.

3. **Avoid twisting.** Hoses are weakened and fittings are loosened by twisting, either during installation or machine operation. Use a hose clamp or allow some free hose where necessary. And remember, tighten the fitting on the hose, not the hose on the fitting.

4. **Avoid rubbing.** Clamp or bracket hoses away from moving parts of sharp edges. If this doesn't work, use a hose guard, wire spring, or flat armor spring design.

5. **Avoid heat.** Keep hoses away from hot surfaces, such as engine manifolds. If you can't route the hoses away from these areas, shield them.

6. **Avoid sharp bends.** The bend radius depends upon the hose construction, size, and pressure. The manufacturer recommends a certain limit for bending on each hose. At lower pressures, a tighter bend is permitted. Where possible, reroute hoses to avoid sharp bends. Or, allow extra slack but watch for kinks or loops.

Hose Fitting Installation Considerations

Besides the rules for installing hoses covered earlier, there are twelve rules for good hose fitting installations, as follows:

1. Be sure the working pressure rating of the coupler corresponds to the working pressure rating of the hose.
2. Be sure the seal replacement is matched to the mating coupler.
3. Use flared adapters or elbow hose fittings where possible, instead of pipe adapters.
4. Improve line routing by using 45° and 90° adapters or elbows.
5. Attach male ends of hose assemblies before female ends.
6. Tighten swivel nuts only until snug—do no overtighten.
7. Tighten only the nipple hex nut and not the socket.
8. Use pipe sealing compound on male threads only—make sure compound is compatible with the hydraulic oil.
9. Use open-end wrenches for assembly—do not use pipe wrenches.
10. Use two wrenches where necessary to prevent twisting of hoses.
11. Tighten the fitting on the hose—not the hose on the fittings.
12. As a general rule, tighten the fitting until finger tight, then use wrenches to tighten the fitting two extra flats. If leakage occurs after operation, tighten one extra flat.

To assemble screw-together-type couplings, place the socket in a vise and turn the hose into the socket until it bottoms. Then back it off ¼ to ½ turn. Next, lubricate the nipple and the inside of the hose with heavy oil. Then thread the nipple into the socket, leaving a small gap. Always make certain the lubricating oil is compatible with the tube. Some couplings require a mandrel for assembly. It is threaded to the nipple and prevents the end of the nipple from damaging the inner tube during assembly. Some couplings are designed to bite directly into the wire reinforcement. For these, the cover must be skived or peeled off. The notch or knurl on the coupling can be used as a guide for the skiving length. To skive a hose, first mark the length, then place it in a vise and cut through the cover with a knife, or with a hacksaw, using backstrokes. Make a diagonal cut and peel the cover off with pliers. Be sure not to damage the reinforcement.

To assemble clamp-type couplings, apply heavy oil to the nipple and the inside of the hose, then insert the nipple into the hose and bolt the segments together over the hose. Be sure to tighten the bolts evenly.

While crimped assemblies are often assembled by the manufacturer, field crimpers allow hose assemblies to be made up on the job site (Figure 6.18).

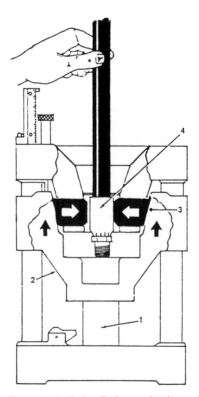

Figure 6.18. Hydraulic hose-crimping unit.

Hose Assembly Failure

Everyone in maintenance encounters hose failures. Normally, there is no problem. The hose is replaced and the equipment goes back on line. Occasionally, though, the failures come too frequently—the same problems keep occurring with the same equipment. At this point, the task of the maintenance person is to determine and correct the cause of these repeated failures.

Failure to look into the problem, if the fault lies with the hose, will simply result in the repeated loss of hose lines. If the problem lies with the equipment, failure to determine the cause could eventually result in the loss of the equipment. A little effort in the beginning can avoid a big headache in the long run.

Every failure should be analyzed, even if that analysis is as basic as deciding that the failure was normal and acceptable. In that case, simply replace the hose line.

However, if the failure rate is unacceptable, probe a little deeper to determine the cause of the failure and correct the situation.

Hose failures fall into five major categories:

1. Improper application
2. Improper assembly and installation
3. External damage
4. Faulty equipment
5. Faulty hose

Improper Application

In order to investigate the most common cause of hose failures—improper application—compare the hose specifications with the requirements of the application. The following areas must be considered:

- The maximum operating pressure of the hose.
- The recommended temperature range of the hose.
- The fluid compatibility of the hose.

Check all of these areas against the requirements of the application. If they do not match up, another type of hose must be selected.

When contacting a distributor or manufacturer for help, provide all of the information needed to solve the problem. Use the following checklist when describing the problem:

1. Either send in a sample of the hose or provide the part number and date of manufacture. This information is supplied on the lay line of the hose.
2. Describe the type of equipment on which the hose is used and the location of the hose on this equipment.
3. Provide the brand name and type of the fluid used with the hose.
4. Give the maximum and minimum temperature, both internal and external, at which the hose operates. Remember that temperatures can vary widely from one part of the equipment to another. Try to get a reading as close to the failed hose as possible.
5. If the hose is bent, provide the bend radius along the inside of the curve or send along a tracing of the curve on a piece of paper. If the hose bends in more than one plane, say so.
6. Provide the flow rate (gpm) through the hose.
7. List the maximum pressure, both static and transient, to which the hose is subjected.
8. Describe the environment in which the hose operates.

By providing complete information, better and faster answers will result.

Improper Assembly and Installation

The second major cause of premature hose failure is improper assembly and installation. This can involve anything from using the wrong fitting on a hose, to poor routing of the hose.

The solution to this problem is, of course, to become thoroughly familiar with the proper assembly and installation of hoses and fittings before installation.

External Damage

External damage can range from abrasion and corrosion, to the hose that is crushed by a lift truck. These are the types of problems that can normally be solved simply once the cause is identified. The hose can be rerouted or clamped, or a fire sleeve or abrasion guard can be used.

In the case of corrosion, the answer may be as simple as changing to a hose with a more corrosion-resistant cover or rerouting the hose to avoid the corrosive element.

Faulty Equipment

Too frequent or premature hose failure can be the symptom of a malfunction in your equipment. This is a factor that should be considered since prompt corrective action can sometimes avoid serious and costly equipment breakdown. The corrective action here would be to review the operation of the equipment to establish that all is in proper order.

Faulty Hose

Occasionally, a failure problem will lie in the hose itself. The most likely cause of a faulty hose is old age. Check the lay line on the hose to determine the date of manufacture. The hose may have exceeded its recommended shelf life.

If it is suspected that the problem lies in the manufacture of the hose, contact the supplier.

Analyzing Hose Failures

A physical examination of the failed hose can often provide a clue to the cause of the failure. Following is a list of symptoms, along with the conditions that could cause them:

1. *Symptom:* The hose inner liner is very hard and has cracked. *Cause:* Heat has a tendency to leach the plasticizers out of the tube. This is a material that gives the hose its flexibility or plasticity.

   Aerated oil causes oxidation to occur in the inner tube. The reaction of oxygen on a rubber product will cause it to harden. Any combination of oxygen and heat will greatly accelerate the hardening of the inner liner. Cavitation occurring inside the inner liner would have the same effect.

2. *Symptom:* The hose has burst, but there is no indication of multiple broken wires along the entire length of the hose. The hose may have burst in more than one place. *Cause:* This would indicate that the pressure has exceeded the minimum burst strength of the hose. Either a stronger hose is needed, or the hydraulic circuit has a malfunction that is causing unusually high-pressure conditions.

3. *Symptom:* Hose has burst on the outside bend and appears to be elliptical in the bent section. In the case of a pump supply

line, the pump is noisy and very hot. The exhaust line on the pump is hard and brittle.

Cause: Violation of the minimum bend radius is most likely the problem in both cases. Check the minimum bend radius and make sure that the application is within specifications. In the case of the pump supply line, partial collapse of the hose is causing the pump to cavitate, creating both noise and heat. This is a most serious situation and will result in catastrophic pump failure if not corrected.

4. *Symptom:* Hose appears to be flattened out in one or two areas and appears to be kinked. It has burst in this area and also appears to be twisted.

   Cause: Torqueing of a hydraulic control hose will tear loose the reinforcement layers and allow the hose to burst through the enlarged gaps between the braided wire strands. Be sure there is never any twisting force on a hydraulic hose.

5. *Symptom:* There are blisters in the outer cover of the hose. If you puncture the blisters, you will find oil in them.

   Cause: A minute pinhole in the inner liner is allowing the high-pressure oil to seep between it and the outer cover. Eventually, it will form a blister wherever the cover adhesion is weakest. In the case of a screw-type reusable fitting, insufficient lubrication of the hose and fitting can cause this condition because the dry inner liner will adhere to the rotating nipple and tear enough to allow seepage. Faulty hose can also cause this condition.

6. *Symptom:* Fitting separated from the end of the hose.

   Cause: It may be that the wrong fitting has been put on the hose. Recheck the manufacturer's specifications and part numbers. In the case of a crimped fitting, the wrong machine setting may have been used, resulting in overcrimping or undercrimping. The outer socket of a screw-together fitting for multiple wire-braided hose may be worn beyond its tolerance. Generally, these sockets should be discarded after being reused about six times. The swaging dies in a swaged-hose assembly may be worn beyond tolerance. The fitting may have been applied improperly to the hose. The hose may have been installed without leaving enough slack to compensate for the possible 4% shortening that may occur when the hose is pressurized. This will impose a great force on the fitting. The hose itself may be out of tolerance.

7. *Symptom:* The inner liner of the hose is badly deteriorated, with evidences of extreme swelling.
   Cause: Indications are that the hose inner liner is not compatible with the fluid being carried. Consult the hose supplier for a compatibility list, or present a sample of the fluid being conducted by the hose for analysis. Be sure that the operating temperatures, both internal and external, do not exceed recommendations.

8. *Symptom:* Hose has burst. The hose cover is badly deteriorated and the surface of the rubber is crazed.
   Cause: This could be caused by old age. The crazed appearance is the effect of weathering and ozone over a period of time. Try to determine the age of the hose.

9. *Symptom:* A spiral reinforced hose has burst and split open, with the wire exploded out and badly entangled.
   Cause: The hose is too short to accommodate the change in length occurring while it is pressurized.

10. *Symptom:* The hose fitting has been pulled out of the hose. The hose has been considerably stretched out in length.
    Cause: Insufficient support of the hose. It is necessary to support a very long length of hose, especially if it is vertical. In these cases, the weight of the hose, along with the weight of the fluid inside the hose, is being imposed on the hose fitting. This force can be transmitted to a wire rope or chain by clamping the hose to it.

11. *Symptom:* Hose is badly flattened out in the burst area. The inner tube is very hard downstream of the burst but appears normal upstream of the burst.
    Cause: The hose has been kinked either by bending it too sharply or by squashing it in some way so that a major restriction was created. As the velocity of the fluid increases through the restriction, the pressure decreases to the vaporization point of the fluid. This condition causes heat and rapid oxidation to take place, which hardens the inner liner of the hose downstream of the restriction.

Troubleshooting Guide and Maintenance Hints

A hydraulic system is quite easy to maintain since the fluid provides a lubricant and protects against overload. But like any other mechanism, it must be operated properly. Hydraulic systems can easily be damaged by too much speed, too much heat, too much pressure, or too much contamination. The key maintenance problems are:

1. Not enough oil in the reservoir
2. Clogged or dirty filters
3. Loose or damaged intake lines
4. Incorrect oil in the system.

All of these problems can be solved or prevented by knowing the system and maintaining it properly.

Knowing the System

Probably the greatest aid to maintaining and troubleshooting a hydraulic system is the confidence of knowing the system. Every component has a purpose. The construction and operating characteristics of each one should be understood. For example, knowing that a solenoid-controlled directional valve can be manually actuated will save considerable time in isolating a defective solenoid. Some additional practices that will increase your troubleshooting abilities, as well as the useful life of the system, include the following:

1. Know the capabilities of the system. Each component in the system has a maximum rated speed, torque, or pressure. Loading the system beyond the specifications simply increases the possibility of failure.
2. Know the correct operating pressures. Always set and check pressures. How else can you know if the operating pressure is above the maximum rating of the components? The question of

the correct operating pressure may arise. If it isn't correctly specified on the hydraulic schematic, the following rule should be applied:

The correct operating pressure is the lowest pressure that will allow adequate performance of the system function and still remain below the maximum rating of the components and machine. Once the correct pressures have been established, note them on the hydraulic schematic for future reference.

3. Know the proper speeds, temperatures, linkage settings, etc. If they aren't specified, check them when the system is functioning correctly and mark them on the schematic for future reference.

Maintenance Hints

Three simple maintenance procedures have the greatest effect on hydraulic-system performance, efficiency, and life. Yet, the very simplicity of them may be the reason they are so often overlooked. (Figure 7.1)

1. Maintaining a clean, sufficient quantity of hydraulic fluid of the proper type and viscosity
2. Changing filters and cleaning strainers
3. Keeping all connections tight, but not to the point of distortion, so that air is excluded from the system

System Troubleshooting

Developing Systematic Procedures

Analyze the system and develop a logical sequence for setting valves, mechanical stops, interlocks, and electrical controls. Tracing flow paths can often be done by listening for flow in the lines or feeling them for warmth. Develop a cause-and-effect troubleshooting guide similar to the charts on pages 262–271.

Recognizing Trouble

The ability to recognize trouble in a specific system is usually acquired with experience. However, a few general trouble indicators can be discussed:

1. Excessive heat means trouble. A misaligned coupling places an excessive load on bearings and can be readily identified by the

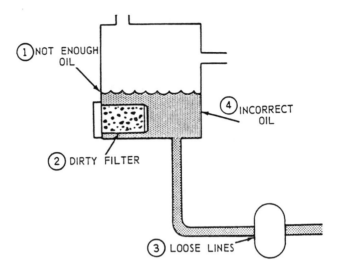

- Keep the oil clean

- Keep the system clean

- Keep your work area clean

- Be careful when you change or add oil

Figure 7.1. Key maintenance problems.

heat generated. A warmer-than-normal tank return line on a relief valve indicates operation at the relief-valve setting. Hydraulic fluids that have a low viscosity will increase the internal leakage of components, resulting in a heat rise. Cavitation and slippage in a pump will also generate heat.

2. Excessive noise means wear, misalignment, or cavitation due to air in the fluid. Contaminated fluid can cause a relief valve to stick and chatter. These noises may be the result of dirty filters, dirty fluid, high-fluid viscosity, excessive drive speed, low-reservoir level, loose intake lines, or worn couplings.

Troubleshooting Procedure

A good troubleshooting and testing program consists of seven basic steps (Figure 7.2):

1. Know how the system works
2. Discuss problems with the machine's operator
3. Operate the machine
4. Inspect the machine
5. Indentify the possible causes of the problem
6. Reach a conclusion as to the diagnostic or corrective action to be taken
7. Test the conclusion

In order to know how the system works, it is necessary to study the machine's technical manuals and service bulletins. The valve settings and pump output should be known, for example.

Discussions should be held with the machine operator to determine how the machine acted, what was unusual, whether operations deteriorated slowly or suddenly, etc. Also, try to find out if any do-it-yourself service was performed.

If it is safe, operate the machine to determine whether gauge readings are proper/normal, if the machine is slow or sluggish, erratic, noisy, and so forth.

Next, the machine should be checked over thoroughly. Inspect the oil in the reservoir to see if it is foamy, milky, has a burnt odor, appears too thin or too thick, etc. Also, check the filters for clogging, visually check the plumbing for any apparent damage or leaks, check the oil cooler, if one is used, and so forth.

All of these steps are intended to collect information regarding the machine condition, behavior, and operation. This data can now be coupled with the knowledge of how the system works to arrive at some possible causes for the problems. These possible causes must be put in an order of priority as to which one is most likely to cause the problem and is easiest to verify. The leading causes for the problem should, of course, be checked first.

The final step is to test the conclusions reached during the analysis of the problem by testing and/or repairing the system. After the repairs are done, check the whole system for leaks, proper oil level, overheatings, pressure settings, operating speed, etc. (Figure 7.3).

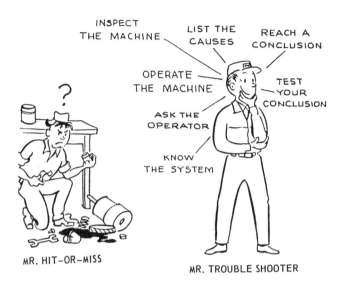

1. **Know the System**
2. **Ask the Operator**
3. **Operate the Machine**
4. **Inspect the Machine**
5. **List the Possible Causes**
6. **Reach a Conclusion**
7. **Test Your Conclusion**

Figure 7.2. Basic troubleshooting steps.

Troubleshooting Guides

The following charts are arranged into five main categories. Each chart highlights a problem that might occur in a hydraulic system. Refer to the proper chart to aid in problem diagnosis and solutions.

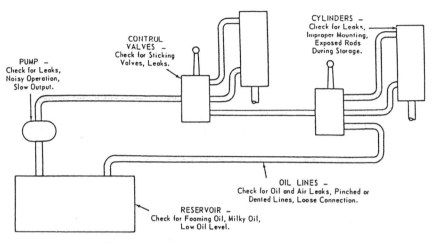

Figure 7.3. Check system before operation.

Diagnosing Hydraulic System Problems

Chart I
Symptom: Excessive Noise

Pump Noisy		Motor Noisy		Relief Valve Noisy	
Cause	Remedy	Cause	Remedy	Cause	Remedy
Cavitation	a	Coupling misaligned	c	Setting too low or too close to another setting	d
Air in fluid	b	Motor or coupling worn or damaged	e	Worn poppet and seat	e
Coupling misaligned	c				
Pump worn or damaged	e				

Remedies

a. Any or all of the following:
- Replace dirty filters.
- Clean clogged inlet line.
- Clean reservoir breather vent.
- Change system fluid.

b. Any or all of the following:
   - Tighten leaky inlet connections.
   - Fill reservoir to proper level.
   - Bleed air from system.
   - Replace pump-shaft seal.
c. Align unit and check condition of seals, bearings, and couplings.
d. Install pressure gauge and adjust to correct pressure.
e. Overhaul or replace.

Chart II
Symptom: Excessive Heat

Pump Heated		Motor Heated		Relief Valve Heated		Fluid Heated	
Cause	Remedy	Cause	Remedy	Cause	Remedy	Cause	Remedy
Fluid heated	d	Fluid heated	d	Fluid heated	d	System pressure too high	d
Cavitation	a	Relief or unloading valve set too high	d	Valve setting incorrect	d	Unloading valve set too high	d
Air in fluid	b	Excessive load	c	Worn or damaged valve	e	Fluid dirty or low supply	f
Relief or unloading valve set too high	d	Worn or damaged motor	e			Incorrect fluid viscosity	f
Excessive load	c					Faulty fluid cooling system	g

264

Worn or damaged pump				Worn pump, valve, motor, cylinder, or other component
e				e

Remedies:

a. Any or all of the following:
  - Replace dirty filters.
  - Clean clogged inlet line.
  - Clean reservoir breather vent.
  - Change system fluid.
  - Change to proper pump drive motor speed.

b. Any or all of the following:
  - Tighten leaky inlet connections.
  - Fill reservoir to proper level.
  - Bleed air from system.
  - Replace pump-shaft seal.

c. Align unit and check condition of seals and bearings. Locate and correct mechanical binding.

d. Install pressure gauge and adjust to correct pressure.

e. Repair or replace part.

f. Change filters and system fluid if of improper viscosity. Fill reservoir to proper level.

g. Clean cooler and/or cooler strainer. Replace cooler control valve. Repair or replace cooler.

Chart III
Symptom: Incorrect Flow

No Flow		Low Flow		Excessive Flow	
Cause	Remedy	Cause	Remedy	Cause	Remedy
Pump not receiving fluid	a	Flow control set too low	d	Flow control set too high	d
Pump-drive motor not operating	e	Relief or unloading valve set too low	d	Yoke actuating device inoperative (variable displacement pumps)	e
Pump-to-drive coupling sheared	e	Flow by-passing thru partially open valve	e or f	Rpm of pump-drive motor incorrect	h
Pump-drive motor turning in wrong position	g	External leak in system	b	Improper size pump used for replacement	h
Directional control set in wrong position	f	Yoke-actuating device inoperative (variable displacement pumps)	e		
Entire flow passing over relief valve	d	Rpm of pump-drive motor incorrect	h		

Damaged pump	c		e
Worn pump, valve, motor, cylinder, or other component	c	e	
Improperly assembled pump	e		

Remedies:

a. Any or all of the following:
   - Replace dirty filters.
   - Clean clogged inlet line.
   - Clean reservoir breather vent.
   - Fill reservoir to proper level.
b. Tighten leaky connections. Bleed air from system.
c. Check for damaged pump or pump drive. Replace and align coupling.
d. Adjust setting.
e. Repair or replace part.
f. Check position of manually operated controls. Check electrical circuit on solenoid-operated controls.
g. Reverse rotation.
h. Replace with correct unit.

Chart IV
Symptom: Incorrect Pressure

No Pressure		Low Pressure			Erratic Pressure		Excessive Pressure	
Cause	Remedy	Cause	Remedy		Cause	Remedy	Cause	Remedy
No flow	See Chart III, first column	Pressure-relief path exists	See Chart III, first & second columns		Air in fluid	b	Pressure-reducing, relief, or unloading valve misadjusted	c
		Pressure-reducing valve set too low	c		Worn relief valve	d	Yoke-actuator inoperative (variable-displacement pumps)	
		Pressure-reducing valve damaged	d		Contamination in fluid	a	Pressure-reducing, relief, or unloading valve worn or damaged	d

	Damaged pump, motor, or cylinder	d	Worn pump, motor, or cylinder	d		

Remedies:

a. Replace dirty filters and system fluid.
b. Tighten leaky connections (fill reservoir to proper level and bleed air from system).
c. Adjust setting.
d. Repair or replace part.

Chart V
Symptom: Faulty Operation

No Movement		Slow Movement		Erratic Movement		Excessive Speed or Movement	
Cause	Remedy	Cause	Remedy	Cause	Remedy	Cause	Remedy
No flow or pressure	See Chart III	Low flow	See Chart III	Erratic pressure	See Chart IV	Excessive flow	See Chart III
Limit or sequence device (mechanical, electrical, or hydraulic) inoperative or maladjusted	c	Fluid viscosity too high	a	Air in fluid	See Chart I	Overriding workload	e
Mechanical bind	b	Insufficient control pressure for valves	See Chart IV	No lubrication of machine ways or linkage	d		

Worn or damaged cylinder or motor	c		c
	No lubrication of machine way or linkage	d	Worn or damaged cylinder or motor
	Worn or damaged cylinder or motor	c	

Remedies:

a. Fluid may be too cold or should be changed to clean fluid of correct viscosity.
b. Locate bind and repair.
c. Repair or replace part.
d. Lubricate.
e. Adjust, repair, or replace counterbalance valve or brake valve.

Appendix

AMERICAN NATIONAL STANDARD

GRAPHIC SYMBOLS FOR FLUID

POWER DIAGRAMS

ANSI standard y32.10–1967

1. INTRODUCTION

1.1 General Fluid power systems are those that transmit and control power through use of a pressurized fluid (liquid or gas) within an enclosed circuit.
Types of symbols commonly used in drawing circuit diagrams for fluid power systems are Pictorial, Cutaway. and Graphic. These symbols are fully explained in the American Standards Association Drafting Manual.

1.1.1. *Pictorial symbols* are very useful for showing the interconnection of components. They are difficult to standardize from a functional basis.

1.1.2. *Cutaway symbols* emphasize construction. These symbols are complex to draw and the functions are not readily apparent.

1.1.3. *Graphic symbols* emphasize the function and methods of operation of components. These symbols are simple to draw. Component functions and methods of operation are obvious. Graphic symbols are capable of crossing language barriers, and can promote a universal understanding of fluid power systems.
Graphic symbols for fluid power systems should be used in conjunction with the graphic symbols for other systems published by the ANSI.

1.2 Scope and Purpose

1.2.1 Scope This standard presents a s y s t e m of graphic symbols for fluid power diagrams.

1.2.1.1. Elementary forms of symbols are:

Circles	Triangles	Lines
Squares	Arcs	Dots
Rectangles	Arrows	Crosses

1.2.1.2 Symbols using words or their abbreviations are avoided. Symbols capable of crossing language barriers are presented herein.

1.2.1.3 Component function rather than construction is emphasized by the symbol.

1.2.1.4 The means of operating fluid power components are shown as part of the symbol (where applicable).

1.2.1.5. This standard shows the basic symbols, describes the principles on which the symbols are based, and illustrates some representative composite symbols. Composite symbols can be devised for any fluid power component by combining basic symbols.
Simplified symbols which are composites of basic symbols are shown for commonly used components.

1.2.1.6. This standard provides basic symbols which differentiate between hydraulic and pneumatic fluid power media.

1.2.2. Purpose

1.2.2.1 The purpose of this standard is to provide a system of fluid power graphic symbols for industrial and educational purposes.

1.2.2.2 The purpose of this standard is to simplify design, fabrication, analysis, and service of fluid power circuits.

1.2.2.3 The purpose of this standard is to provide fluid power graphic symbols which are internationally recognized.

1.2.2.4 The purpose of this standard is to promote universal understanding of fluid power systems.

1.3 Terms and Definitions Terms and corresponding definitions found in this standard are listed elsewhere
The notation **IEC** adjacent to symbols indicates that symbol has been recommended by the International Electrotechnical Committee.
The notation **ISO** adjacent to symbols indicates that adoption of that symbol is under consideration by the International Organization for Standardization.

2. SYMBOL RULES

(See Section 10)

2.1 Symbols show connections, flow paths, and functions of components represented. They can indicate conditions occurring during transition from one flow path arrangement to another. Symbols do not indicate construction, nor do they indicate values, such as pressure, flow rate, and other component settings.

2.2 Symbols do not indicate locations of ports, direction of shifting of spools, or positions of control elements on actual component.

2.3 Symbols may be rotated or reversed without altering their meaning except in the cases of: a.) Lines to Reservoir, 4.1.1; b.) Vented Manifold, 4.1.2.3; c.) Accumulator, 4.2.

2.4 Line technique
Keep line widths approximately equal. Line width does not alter meaning of symbols.

2.4.1 ————————— Solid Line
(Main line conductor, outline, and shaft)

2.4.2 —— —— —— —— —— Dash Line
(Pilot line for control)

2.4.3 ———————————— Dotted Line
(Drain line)

2.4.4 ——— — ——— — ——— Center Line
(Enclosure outline)

2.4.5 Lines crossing (The intersection is not necessarily at a 90° angle.)

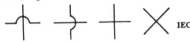 IEC

2.4.6 Lines joining

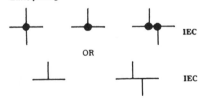 IEC

OR

IEC

2.5 Basic symbols may be shown any suitable size. Size may be varied for emphasis or clarity. Relative sizes should be maintained. (As in the following example.)

2.5.1 Circle and Semi-Circle

2.5.1.1 Large and small circles may be used to signify that one component is the "main" and the other the auxiliary.

2.5.2 Triangle

2.5.3 Arrow ——————➤

2.5.4 Square   Rectangle

2.6 Letter combinations used as parts of graphic symbols are not necessarily abbreviations.

2.7 In multiple envelope symbols, the flow condition shown nearest an actuator symbol takes place when that control is caused or permitted to actuate.

2.8 Each symbol is drawn to show normal, at-rest, or neutral condition of component unless multiple diagrams are furnished showing various phases of circuit operation.

2.9 An arrow through a symbol at approximately 45° indicates that the component can be adjusted or varied.

2.10 An arrow parallel to the short side of a symbol, within the symbol, indicates that the component is pressure compensated.

2.11 A line terminating in a dot to represent a thermometer is the symbol for temperature cause or effect.
See Temperature Controls 7.9, Temperature Indicators and Recorders 9.1.2. and Temperature Compensation 10.16.3 and .4.

2.12 External ports are located where flow lines connect to basic symbol, except where component enclosure symbol is used.
External ports are located at intersections of flow lines and component enclosure symbol when enclosure is used, see SECTION 11.

2.13 Rotating shafts are symbolized by an arrow which indicates direction of rotation (assume arrow on near side of shaft).

3. CONDUCTOR, FLUID

3.1 ──────── Line. Working (main)

3.2 ─ ─ ─ ─ ─ Line, Pilot (for control)

3.3 ─ ─ ─ ─ ─ ─ ─ Line, Liquid Drain

3.4 Line, sensing, etc. such as gauge lines shall be drawn the same as the line to which it connects.

3.5 **Flow, Direction of**

3.5.1 Pneumatic

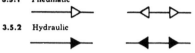

3.5.2 Hydraulic

3.6 **Line, Pneumatic Outlet to Atmosphere**

3.6.1 Plain orifice, unconnectable

3.6.2 Connectable orifice (e.g. Thread) by
Internal Return

3.7 **Line with Fixed Restriction**

3.8 **Line, Flexible**

3.9 **Station, Testing, measurement, or power take-off**

3.91 Plugged port

3.10 **Quick Disconnect**

3.10.1 Without Checks

Connected

Disconnected

3.10.2   With Two Checks

Connected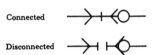

Disconnected

3.10.3   With One Check

Connected

Disconnected

3.11   Rotating Coupling

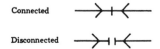

4. ENERGY STORAGE AND FLUID STORAGE

4.1   **Reservoir**

Vented   Pressurized

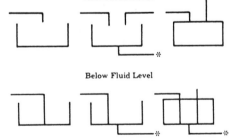

Note: Reservoirs are conventionally drawn in the horizontal plane. All lines enter and leave from above. Examples:

4.1.1   **Reservoir with Connecting Lines**
Above Fluid Level

Below Fluid Level

* Show line entering or leaving below reservoir only when such bottom connection is essential to circuit function.

4.1.2   **Simplified symbol.** The symbols are used as part of a complete circuit. They are analogous to the ground symbol of electrical diagrams. ──│ⁱ IEC Several such symbols may be used in one diagram to represent the same reservoir.

4.1.2.1   Below Fluid Level

4.1.2.2   Above Fluid Level
(The return line is drawn to terminate at the upright legs of the tank symbol.)

4.1.2.3   Vented Manifold

4.2   Accumulator

4.2.1   Accumulator, Spring Loaded

4.2.2   Accumulator, Gas Charged

4.2.3   Accumulator, Weighted

4.3   Receiver, for Air or other Gases

4.4   Energy Source (Pump, Compressor, Accumulator, etc.) This symbol may be used to represent a fluid power source which may be a pump, compressor, or another associated system.

─────▶──── Hydraulic

─────▷──── Pneumatic

Simplified Symbol

Example:

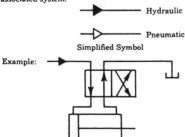

5. FLUID CONDITIONERS

Devices which control the physical characteristics of the fluid.

5.1   Heat Exchanger

5.1.1   Heater
Inside triangles indicate the introduction of heat.

Outside triangles show the heating medium is liquid.

Outside triangles show the heating medium is gaseous.

5.1.2   Cooler

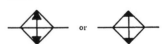

Inside triangles indicate heat dissipation.

(Corners may be filled in to represent triangles.)

5.1.3   Temperature Controller—(The temperature is to be maintained between two predetermined limits.)

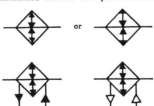

5.2   Filter - Strainer

5.3   Separator

5.3.1   With Manual Drain

5.3.2   With Automatic Drain

5.4   Filter - Separator

5.4.1   With Manual Drain

5.4.2   With Automatic Drain

5.5   Dessicator (Chemical Dryer)

5.6   Lubricator

5.6.1   Less Drain

5.6.2   With Manual Drain

6. LINEAR DEVICES

6.1   Cylinders, Hydraulic & Pneumatic

6.1.1   Single Acting

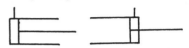

6.1.2   Double Acting

6.1.2.1   Single End Rod

6.1.2.2   Double End Rod

6.1.2.3   Fixed Cushion, Advance & Retract

6.1.2.4   Adjustable Cushion, Advance Only

6.1.2.5   Use these symbols when diameter of rod compared to diameter of bore is significant to circuit function.

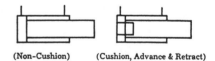

(Non-Cushion)   (Cushion, Advance & Retract)

6.2   Pressure Intensifier

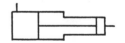

6.3   Servo Positioner (Simplified)
Hydraulic   Pneumatic

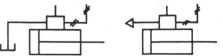

6.4   Discrete Positioner
Combine two or more basic cylinder symbols.

7. ACTUATORS & CONTROLS

7.1   Spring

7.2   Manual
(Use as general symbol without indication of specific type: i.e., foot, hand, leg, arm)

7.2.1   Push Button

7.2.2   Lever

7.2.3   Pedal or Treadle

7.3   Mechanical

7.4   Detent
(Show a notch for each detent in the actual component being symbolized. A short line indicates which detent is in use. Detent may be positioned on symbol for drafting convenience.)

7.5   Pressure Compensated

7.6   Electrical

7.6.1   Solenoid (Single Winding)

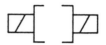

7.6.2   Reversing Motor

7.7   Pilot Pressure

7.7.1
Remote Supply

7.7.2
Internal Supply

7.7.3   Actuation By Released Pressure

Remote Exhaust   Internal Return

7.7.4   Pilot Controlled, Spring Centered

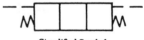

Simplified Symbol

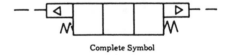

Complete Symbol

7.7.5   Pilot Differential

Simplified Symbol

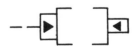

Complete Symbol

7.8   Solenoid Pilot

7.8.1   Solenoid or Pilot

External Pilot
Supply

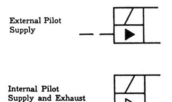

Internal Pilot
Supply and Exhaust

7.8.2   Solenoid and Pilot

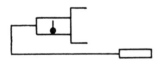

7.9   Thermal—A mechanical device responding to thermal change.

7.9.1   Local Sensing

7.9.2   With Bulb for Remote Sensing

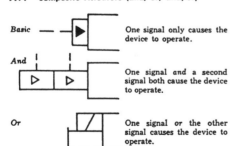

7.10   Servo
(This symbol contains representation for energy input, command input, and resultant output.)

7.11   Composite Actuators (and, or, and/or)

Basic

One signal only causes the device to operate.

And

One signal and a second signal both cause the device to operate.

Or

One signal or the other signal causes the device to operate.

And/Or

The solenoid and the pilot or the manual override alone causes the device to operate.

The solenoid and the pilot or the manual override and the pilot.

The solenoid and the pilot
or
a manual override and the pilot
or
a manual override alone.

8. ROTARY DEVICES

8.1   Basic Symbol

8.1.1   With Ports

8.1.2   With Rotating Shaft, with control, and with Drain

8.2   Hydraulic Pump

8.2.1   Fixed Displacement

8.2.1.1   Unidirectional

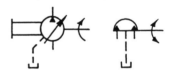

8.2.1.2   Bidirectional

8.2.2   Variable Displacement, Non-Compensated

8.2.2.1   Unidirectional

Simplified   Complete

8.2.2.2   Bidirectional

Simplified   Complete

8.2.3   **Variable Displacement, Pressure Compensated**

8.2.3.1   Unidirectional

Simplified   Complete

8.2.3.2   Bidirectional

Simplified   Complete

8.3   **Hydraulic Motor**

8.3.1   **Fixed Displacement**

8.3.1.2   **Bidirectional**

8.3.2   **Variable Displacement**

8.3.2.1   Unidirectional

8.3.2.2   Bidirectional

8.4   **Pump-Motor, Hydraulic**

8.4.1   Operating in one direction as a pump.
Operating in the other direction as a motor.

8.4.1.1   **Complete Symbol**

8.4.1.2   Simplified Symbol

8.4.2   Operating one direction of flow as either a pump
or as a motor.

8.4.2.1   Complete Symbol

8.4.2.2   Simplified Symbol

8.4.3   Operating in both directions of flow either as a pump or as a motor. (Variable displacement, pressure compensated shown.)

8.4.3.1   Complete Symbol

8.4.3.2   Simplified Symbol

8.5   **Pump, Pneumatic**

8.5.1   **Compressor, Fixed Displacement**

8.5.2   **Vacuum Pump, Fixed Displacement**

8.6   **Motor, Pneumatic**

8.6.1   **Unidirectional**

8.6.2   **Bidirectional**

8.7   **Oscillator**

8.7.1   **Hydraulic**

8.7.2   **Pneumatic**

8.8   **Motors, Engines**

8.8.1   **Electric Motor**

8.8.2   **Heat Engine**
   (Internal Combustion Engine, Steam)

9. INSTRUMENTS & ACCESSORIES

9.1   **Indicating and Recording**

9.1.1   **Pressure**

9.1.2   **Temperature**

9.1.3   **Flow Meter**

9.1.3.1   **Flow Rate**

9.1.3.2   **Totalizing**

9.2   **Sensing**

9.2.1 Venturi

9.2.2 Orifice Plate

9.2.3 Pitot Tube

9.2.4 Nozzle
  Hydraulic   Pneumatic

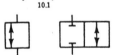

9.3 Accessories

9.3.1 Pressure Switch

9.3.2 Float Switch

9.3.3 Muffler

10. VALVES

A basic valve symbol is composed of one or more envelopes with lines inside the envelope to represent flow paths and flow conditions between ports. Three symbol systems are used to represent valve types: single envelope, both finite and infinite position; multiple envelope, finite position; and multiple envelope, infinite position.

10.1 In finite position single envelope valves, the envelope is imagined to move to illustrate how pressure or flow conditions are controlled as the valve is actuated.

10.2 Multiple envelopes symbolize valves providing more than one finite flow path option for the fluid. The multiple envelope moves to represent how flow paths change when the valving element within the component is shifted to its finite positions.

10.3 Multiple envelope valves capable of infinite positioning between certain limits are symbolized as in 10.2 above with the addition of horizontal bars which are drawn parallel to the envelope. The horizontal bars are the clues to the infinite positioning function possessed by the valve represented.

10.4 Envelopes

10.5 Ports

10.6 Ports, Internally Blocked
    Symbol System   Symbol System
        10.1   10.2

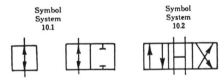

10.7 Flow Paths, Internally Open (Symbol Systems 10.1 & 10.2)

    Symbol   Symbol
    System   System
     10.1   10.2

10.8 Flow Paths, internally Open (Symbol System 10.3)

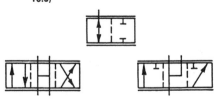

10.9 Two-Way Valves (2 Ported Valves)

10.9.1   On-Off (Manual Shut-Off)

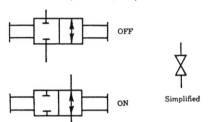

OFF

ON

Simplified

10.9.2   Check

Simplified Symbol

Flow to the right is blocked.
Flow to the left is permitted.
(Composite Symbol)

10.9.3   Check, Pilot-Operated to Open

10.9.4   Check, Pilot-Operated to Close

10.9.5   Two-Way Valves

10.9.5.1   Two-Position
Normally Closed   Normally Open

10.9.5.2   Normally Open

10.9.5.2   Infinite Position
Normally Closed   Normally Open
Complete

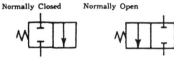

10.10   Three-Way Valves

10.10.1   Two-Position

10.10.1.1   Normally Open

10.10.1.2   Normally Closed

10.10.1.3   Distributor (Pressure is distributed first to one port, then the other)

10.10.1.4   Two-Pressure

10.10.2   Double Check Valve. Double check valves can be built with and without 'cross bleed.' Such valves with two poppets do not usually allow pressure to momentarily 'cross bleed' to return during transition. Valves with one poppet may allow 'cross bleed' as these symbols illustrate.

10.10.2.1   Without Cross Bleed (One Way Flow)

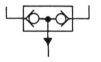

10.10.2.2   With Cross Bleed (Reverse Flow Permitted)

10.11   Four-Way Valves

10.11.1   Two Position

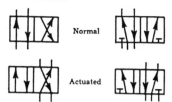

Normal

Actuated

10.11.2   Three Position

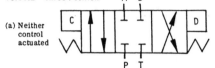

(a) Neither control actuated

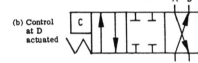

(b) Control at D actuated

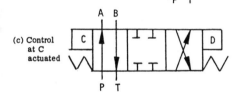

(c) Control at C actuated

10.11.3   Typical Flow Paths for Center Condition of Three-Position Valves

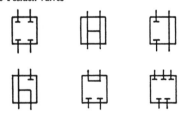

10.11.4   Two-Position, Snap Action with Transition. As the valve element shifts from one position to the other, it passes through an intermediate position. If it is essential to circuit function to symbolize this "in transit" condition, it can be shown in the center position, enclosed by dashed lines.

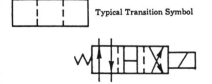

Typical Transition Symbol

10.12   Infinite Positioning (Between Open & Closed)

10.12.1   Normally Closed

10.12.2   Normally Open

10.13   Pressure Control Valves

10.13.1   Pressure Relief

Simplified Symbol Denotes

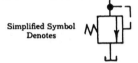

Normal   Actuated (Relieving)

10.13.2   Sequence

10.13.3   Pressure Reducing

10.13.4 Pressure Reducing and Relieving

10.13.5 Airline Pressure Regulator (Adjustable, Relieving)

10.14 Infinite Positioning Three-Way Valves

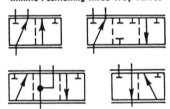

10.15 Infinite Positioning Four-Way Valves

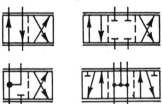

10.16 Flow Control Valves (See 3.7)

10.16.1 Adjustable, Non-Compensated (Flow control in each direction)

10.16.2 Adjustable with Bypass
Flow is controlled to the right.
Flow to the left bypasses control.

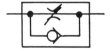

10.16.3 Adjustable and Pressure Compensated with Bypass

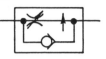

10.16.4 Adjustable, Temperature & Pressure Compensated

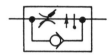

11. REPRESENTATIVE COMPOSITE SYMBOLS

11.1 Component Enclosure. Component enclosure may surround a complete symbol or a group of symbols to represent an assembly. It is used to convey more information about component connections and functions.

Enclosure indicates extremity of component or assembly. External ports are assumed to be on enclosure line and indicate connections to component.
Flow lines shall cross enclosure line without loops or dots.

11.3.1.4 Variable Displacement with Integral Replenishing Pump and Control Valves.

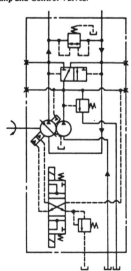

11.3.2   Pump Motor. Variable displacement with manual, electric, pilot, and servo control.

11.4.9   Multiple, Three Position, Manual Directional Control with Integral Check and Relief Valves

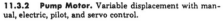

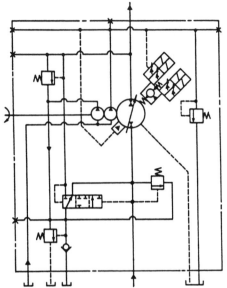

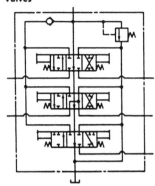

Other representative composites omitted for brevity

DIRECTIONAL CONTROL VALVES

Valves, directional control. Start, stop, and direct fluid flow. They extend and retract cylinders, rotate fluid motors and actuators, and sequence other circuit operations.

Directional control valves are classified according to the number of ports or connecting lines, the number of positions to which they can be actuated, the type of actuator, and the way in which fluid flows through the valve.

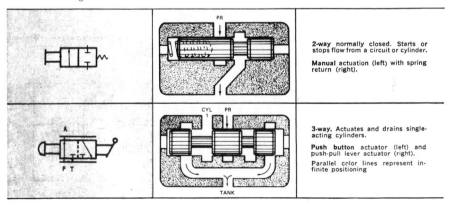

2-way normally closed. Starts or stops flow from a circuit or cylinder.

Manual actuation (left) with spring return (right).

3-way. Actuates and drains single-acting cylinders.

Push button actuator (left) and push-pull lever actuator (right).

Parallel color lines represent infinite positioning

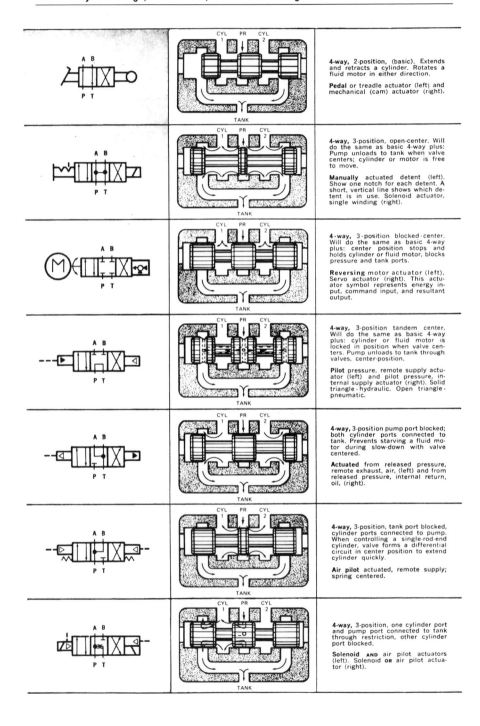

4-way, 2-position, (basic). Extends and retracts a cylinder. Rotates a fluid motor in either direction.

Pedal or treadle actuator (left) and mechanical (cam) actuator (right).

4-way, 3-position, open-center. Will do the same as basic 4-way plus: Pump unloads to tank when valve centers; cylinder or motor is free to move.

Manually actuated detent (left). Show one notch for each detent. A short, vertical line shows which detent is in use. Solenoid actuator, single winding (right).

4-way, 3-position blocked-center. Will do the same as basic 4-way plus: center position stops and holds cylinder or fluid motor, blocks pressure and tank ports.

Reversing motor actuator (left). Servo actuator (right). This actuator symbol represents energy input, command input, and resultant output.

4-way, 3-position tandem center. Will do the same as basic 4-way plus: cylinder or fluid motor is locked in position when valve centers. Pump unloads to tank through valves, center-position.

Pilot pressure, remote supply actuator (left) and pilot pressure, internal supply actuator (right). Solid triangle - hydraulic. Open triangle - pneumatic.

4-way, 3-position pump port blocked; both cylinder ports connected to tank. Prevents starving a fluid motor during slow-down with valve centered.

Actuated from released pressure, remote exhaust, air, (left) and from released pressure, internal return, oil, (right).

4-way, 3-position, tank port blocked, cylinder ports connected to pump. When controlling a single-rod-end cylinder, valve forms a differential circuit in center position to extend cylinder quickly.

Air pilot actuated, remote supply; spring centered.

4-way, 3-position, one cylinder port and pump port connected to tank through restriction, other cylinder port blocked.

Solenoid AND air pilot actuators (left). Solenoid **OR** air pilot actuator (right).

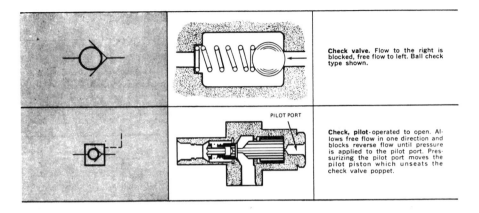

Check valve. Flow to the right is blocked, free flow to left. Ball check type shown.

PILOT PORT

Check, pilot-operated to open. Allows free flow in one direction and blocks reverse flow until pressure is applied to the pilot port. Pressurizing the pilot port moves the pilot piston which unseats the check valve poppet.

FLOW CONTROL VALVES

Valves, flow control. Control flow by restricting fluid movement in one or both directions. The restriction may be fixed or variable. In some elementary flow control valves, flow varies with changes in fluid line pressure. The more sophisticated ones compensate for fluid pressure and temperature fluctuations.

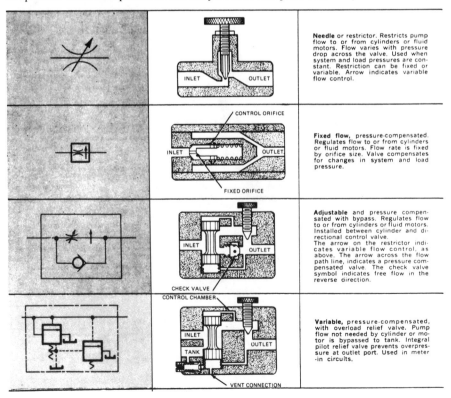

INLET   OUTLET

Needle or restrictor. Restricts pump flow to or from cylinders or fluid motors. Flow varies with pressure drop across the valve. Used when system and load pressures are constant. Restriction can be fixed or variable. Arrow indicates variable flow control.

CONTROL ORIFICE

INLET   OUTLET

FIXED ORIFICE

Fixed flow, pressure-compensated. Regulates flow to or from cylinders or fluid motors. Flow rate is fixed by orifice size. Valve compensates for changes in system and load pressure.

INLET   OUTLET

CHECK VALVE

Adjustable and pressure compensated with bypass. Regulates flow to or from cylinders or fluid motors. Installed between cylinder and directional control valve. The arrow on the restrictor indicates variable flow control, as above. The arrow across the flow path line, indicates a pressure compensated valve. The check valve symbol indicates free flow in the reverse direction.

CONTROL CHAMBER

INLET   OUTLET

TANK

VENT CONNECTION

Variable, pressure-compensated, with overload relief valve. Pump flow not needed by cylinder or motor is bypassed to tank. Integral pilot relief valve prevents overpressure at outlet port. Used in meter-in circuits.

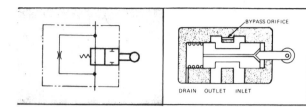

Deceleration. Meters flow from a cylinder near the end of its stroke to decelerate the load. May have integral check valve for free reverse flow. Shaft of actuating cam determines rate of deceleration.

PRESSURE CONTROL VALVES

Valves, pressure control. Limit system pressure, reduce pressure in any part of a circuit, unload a pump during part of a machine's cycle, or determine the pressure at which oil enters part of a circuit. Depending on their function in a circuit, they are called relief, pressure reducing, unloading, sequence, or counterbalance.

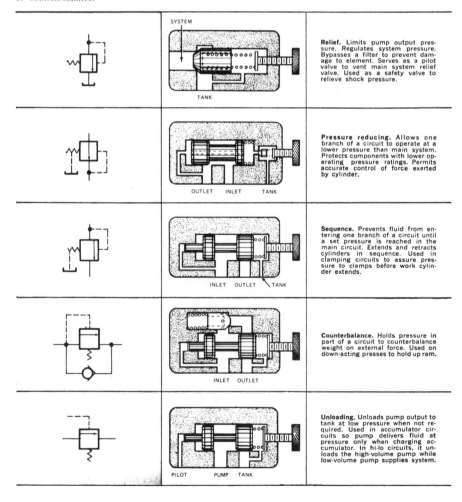

Relief. Limits pump output pressure. Regulates system pressure. Bypasses a filter to prevent damage to element. Serves as a pilot valve to vent main system relief valve. Used as a safety valve to relieve shock pressure.

Pressure reducing. Allows one branch of a circuit to operate at a lower pressure than main system. Protects components with lower operating pressure ratings. Permits accurate control of force exerted by cylinder.

Sequence. Prevents fluid from entering one branch of a circuit until a set pressure is reached in the main circuit. Extends and retracts cylinders in sequence. Used in clamping circuits to assure pressure to clamps before work cylinder extends.

Counterbalance. Holds pressure in part of a circuit to counterbalance weight on external force. Used on down-acting presses to hold up ram.

Unloading. Unloads pump output to tank at low pressure when not required. Used in accumulator circuits so pump delivers fluid at pressure only when charging accumulator. In hi-lo circuits, it unloads the high-volume pump while low-volume pump supplies system.

LINEAR ACTUATORS

These convert fluid energy into linear mechanical force and motion. They usually consist of a movable element such as a piston and piston rod operating within a close-fitting cylindrical bore.

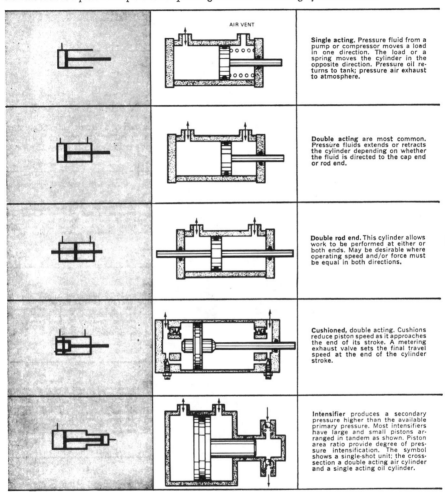

Single acting. Pressure fluid from a pump or compressor moves a load in one direction. The load or a spring moves the cylinder in the opposite direction. Pressure oil returns to tank; pressure air exhaust to atmosphere.

Double acting are most common. Pressure fluids extends or retracts the cylinder depending on whether the fluid is directed to the cap end or rod end.

Double rod end. This cylinder allows work to be performed at either or both ends. May be desirable where operating speed and/or force must be equal in both directions.

Cushioned, double acting. Cushions reduce piston speed as it approaches the end of its stroke. A metering exhaust valve sets the final travel speed at the end of the cylinder stroke.

Intensifier produces a secondary pressure higher than the available primary pressure. Most intensifiers have large and small pistons arranged in tandem as shown. Piston area ratio provide degree of pressure intensification. The symbol shows a single-shot unit; the cross-section a double acting air cylinder and a single acting oil cylinder.

PUMPS

Converts mechanical force and motion to hydraulic power.

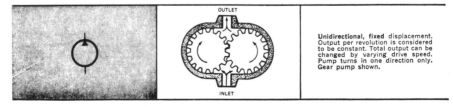

Unidirectional, fixed displacement. Output per revolution is considered to be constant. Total output can be changed by varying drive speed. Pump turns in one direction only. Gear pump shown.

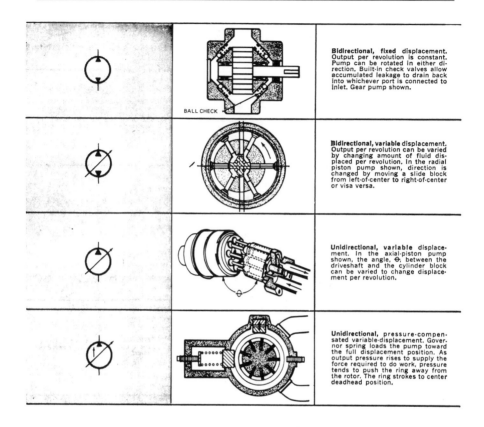

	BALL CHECK	**Bidirectional, fixed displacement.** Output per revolution is constant. Pump can be rotated in either direction. Built-in check valves allow accumulated leakage to drain back into whichever port is connected to inlet. Gear pump shown.
		Bidirectional, variable displacement. Output per revolution can be varied by changing amount of fluid displaced per revolution. In the radial piston pump shown, direction is changed by moving a slide block from left-of-center to right-of-center or visa versa.
		Unidirectional, variable displacement. In the axial-piston pump shown, the angle, Θ, between the driveshaft and the cylinder block can be varied to change displacement per revolution.
		Unidirectional, pressure-compensated variable-displacement. Governor spring loads the pump toward the full displacement position. As output pressure rises to supply the force required to do work, pressure tends to push the ring away from the rotor. The ring strokes to center deadhead position.

MOTORS

Converts fluid power into mechanical force and motion.

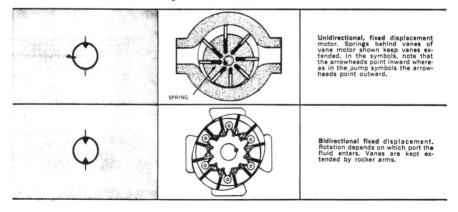

	SPRING	**Unidirectional, fixed displacement** motor. Springs behind vanes of vane motor shown keep vanes extended. In the symbols, note that the arrowheads point inward whereas in the pump symbols the arrowheads point outward.
		Bidirectional fixed displacement. Rotation depends on which port the fluid enters. Vanes are kept extended by rocker arms.

COMPRESSORS

A machine that converts mechanical energy to pneumatic energy. Takes air at atmospheric pressure and compresses it to higher pressure.

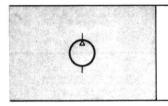

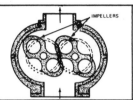

Rotary compressor. A pair of external gears keep two lobes in proper rotative relationship. No lubrication is needed on lobe surfaces because there is no internal contact pressure.

PRESSURE SWITCH

Fluid pressure actuates an electric switch. Pressure switch can sense and react to flow, liquid level or velocity.

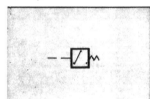

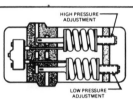

HIGH PRESSURE ADJUSTMENT

LOW PRESSURE ADJUSTMENT

Differential pressure switch. Has a low pressure and a high pressure adjustment.

QUICK DISCONNECT COUPLINGS

Make it possible to separate lines without losing fluid. This eliminates the need for draining a line before disconnecting it. Provide immediate separation or connection with little effort.

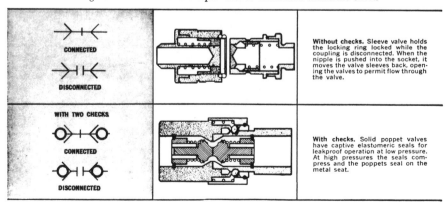

CONNECTED

DISCONNECTED

Without checks. Sleeve valve holds the locking ring locked while the coupling is disconnected. When the nipple is pushed into the socket, it moves the valve sleeves back, opening the valves to permit flow through the valve.

WITH TWO CHECKS

CONNECTED

DISCONNECTED

With checks. Solid poppet valves have captive elastomeric seals for leakproof operation at low pressure. At high pressures the seals compress and the poppets seal on the metal seat.

ROTARY ACTUATOR

Rotates an output shaft through a fixed arc to produce oscillating power. Converts fluid-energy input to mechanical output. Available in a variety of sizes and design types.

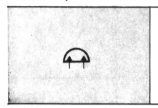 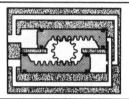

Double rack and pinion. Forces rotating the output shaft are balanced. Moving parts operate in a sealed, oil-filled chamber. Angle of rotation is limited. Usually operates at relatively slow speeds.

ACCUMULATOR

A container in which fluid is stored under pressure as a source of pressure fluid. In a hydropneumatic accumulator, compressed gas applies force to the stored fluid. Mechanical accumulators incorporate a mechanical device which applies the force to the stored fluid.

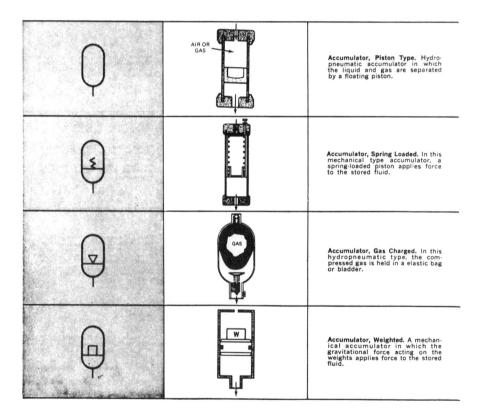

Accumulator, Piston Type. Hydropneumatic accumulator in which the liquid and gas are separated by a floating piston.

Accumulator, Spring Loaded. In this mechanical type accumulator, a spring-loaded piston applies force to the stored fluid.

Accumulator, Gas Charged. In this hydropneumatic type, the compressed gas is held in a elastic bag or bladder.

Accumulator, Weighted. A mechanical accumulator in which the gravitational force acting on the weights applies force to the stored fluid.

FLUID CONDITIONERS

Devices which control the physical characteristics of the fluid. Individual components condition the fluid to meet specific operating needs.

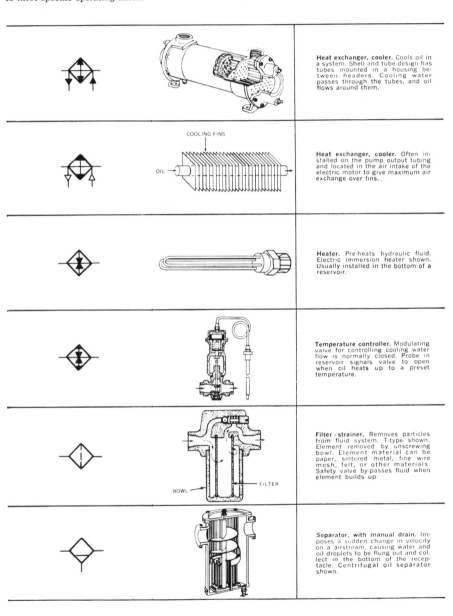

Heat exchanger, cooler. Cools oil in a system. Shell and tube design has tubes mounted in a housing between headers. Cooling water passes through the tubes, and oil flows around them.

Heat exchanger, cooler. Often installed on the pump output tubing and located in the air intake of the electric motor to give maximum air exchange over fins.

Heater. Pre-heats hydraulic fluid. Electric immersion heater shown. Usually installed in the bottom of a reservoir.

Temperature controller. Modulating valve for controlling cooling water flow is normally closed. Probe in reservoir signals valve to open when oil heats up to a preset temperature.

Filter - strainer. Removes particles from fluid system. T-type shown. Element removed by unscrewing bowl. Element material can be paper, sintered metal, fine wire mesh, felt, or other materials. Safety valve by-passes fluid when element builds up.

Separator, with manual drain. Imposes a sudden change in velocity on a airstream, causing water and oil droplets to be flung out and collect in the bottom of the receptacle. Centrifugal oil separator shown.

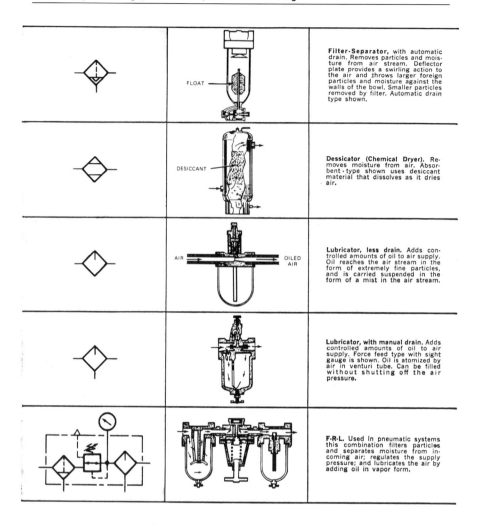

	FLOAT	**Filter-Separator,** with automatic drain. Removes particles and moisture from air stream. Deflector plate provides a swirling action to the air and throws larger foreign particles and moisture against the walls of the bowl. Smaller particles removed by filter. Automatic drain type shown.
	DESICCANT	**Dessicator (Chemical Dryer).** Removes moisture from air. Absorbent-type shown uses desiccant material that dissolves as it dries air.
	AIR / OILED AIR	**Lubricator, less drain.** Adds controlled amounts of oil to air supply. Oil reaches the air stream in the form of extremely fine particles, and is carried suspended in the form of a mist in the air stream.
		Lubricator, with manual drain. Adds controlled amounts of oil to air supply. Force feed type with sight gauge is shown. Oil is atomized by air in venturi tube. Can be filled without shutting off the air pressure.
		F-R-L. Used in pneumatic systems this combination filters particles and separates moisture from incoming air; regulates the supply pressure; and lubricates the air by adding oil in vapor form.

ROTATING COUPLING

Can turn in one, two, or three planes. Leakproof, and have low torque. Help relieve stresses, caused by shock, misalignment, or vibration.

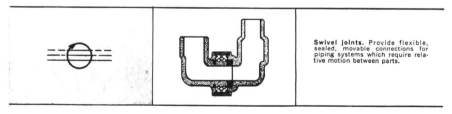

		Swivel joints. Provide flexible, sealed, movable connections for piping systems which require relative motion between parts.

GAGES

An instrument or device for measuring, indicating, or comparing a physical characteristic.

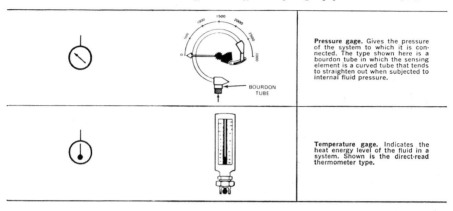

Pressure gage. Gives the pressure of the system to which it is connected. The type shown here is a bourdon tube in which the sensing element is a curved tube that tends to straighten out when subjected to internal fluid pressure.

Temperature gage. Indicates the heat energy level of the fluid in a system. Shown is the direct-read thermometer type.

MUFFLER

A device used in pneumatic systems to reduce air noise and blast. Reduce air velocity, change its direction and reduce vibration. Noise is decreased by back pressure control of gas expansion.

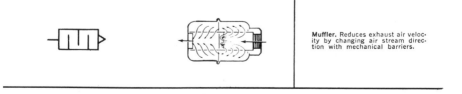

Muffler. Reduces exhaust air velocity by changing air stream direction with mechanical barriers.

Glossary

ACCUMULATOR: A container in which fluid is stored under pressure as a source of fluid power.

ADDITIVE: A chemical compound or compounds added to a fluid to change its properties.

AIR, COMPRESSED: Air at any pressure greater than atmospheric pressure.

AIR, FREE: Air under the pressure due to atmospheric conditions at any specific location.

AIR, STANDARD: Air at a temperature of 68°F, a pressure of 14.70 pounds per square inch absolute, and a relative humidity of 36% (0.0750 pounds per cubic foot). In gas industries the temperature of "standard air" is usually given at 60°F.

AIR BREATHER: A device permitting air movement between atmosphere and the component in which it is installed.

AIR RECEIVER: A container in which gas is stored under pressure as a source of pneumatic fluid power.

ANILINE POINT: The lowest temperature at which a liquid is completely miscible with an equal volume of freshly distilled aniline (ASTM Designation D611-55T).

ANTI-EXTRUSION RING: A ring which bridges a clearance to minimize seal extrusion.

BERNOULLI'S LAW: If no work is done on or by a flowing frictionless liquid, its energy due to pressure and velocity remains constant at all points along the streamline.

BLEEDER: A device for removal of pressurized fluid.

BOYLE'S LAW: The absolute pressure of a fixed mass of gas varies inversely as the volume, provided the temperature remains constant.

BULK MODULUS: The measure of resistance to compressibility of a fluid. It is the reciprocal of the compressibility.

CAP: A cylinder end closure which completely covers the bore area.

CAVITATION: A localized gaseous condition within a liquid stream which occurs where the pressure is reduced to the vapor pressure.

CHARLES' LAW: The volume of a fixed mass of gas varies directly with absolute temperature, provided the pressure remains constant.

CIRCUIT, PILOT: A circuit used to control a main circuit or component.

CIRCUIT, REGENERATIVE: A circuit in which pressurized fluid discharged from a component is returned to the system to reduce power input requirements. On single rod end cylinders the discharge from the rod end is often directed to the opposite end to increase rod extension speed.

CIRCUIT, SERVO: A circuit which is controlled by automatic feed back; i.e., the output of the system is sensed or measured and is compared with the input signal. The difference (error) between the actual output and the input controls the circuit. The controls attempt to minimize the error. The system output may be position, velocity, force, pressure, level, flow rate, or temperature.

CIRCUIT, UNLOADING: A circuit in which pump volume is returned to reservoir at near zero gage pressure whenever delivery to the system is not required.

COMPRESSIBILITY: The change in volume of a unit volume of a fluid when subjected to a unit change of pressure.

COMPRESSOR: A device which converts mechanical force and motion into pneumatic fluid power.

COMPRESSOR, MULTIPLE STAGE: A compressor having two or more compressive steps in which the discharge from each supplies the next in series.

CONTAMINANT: Detrimental matter in a fluid.

CONTROL, PRESSURE COMPENSATED: A control in which a pressure signal operates a compensating device.

COUPLING, QUICK DISCONNECT: A coupling which can quickly join or separate lines.

CROSS: A connector with four ports arranged in pairs, each pair on one axis, and the axes at right angles.

CUSHION, CYLINDER: A cushion built into a cylinder to restrict flow at the outlet port thereby arresting the motion of the piston rod.

CYCLE: A single complete operation consisting of progressive phases starting and ending at the neutral position.

CYLINDER: A device which converts fluid power into linear mechanical force and motion. It usually consists of a movable element such as a piston and piston rod, plunger rod, plunger or ram, operating within a cylindrical bore.

CYLINDER, DOUBLE ACTING: A cylinder in which fluid force can be applied to the movable element in either direction.

CYLINDER, PISTON: A cylinder in which the movable element has a greater cross-sectional area than the piston rod.

CYLINDER, PLUNGER: A cylinder in which the movable element has the same cross-sectional area as the piston rod.

CYLINDER, SINGLE ACTING: A cylinder in which the fluid force can be applied to the movable element in only one direction.

CYLINDER, TANDEM: Two or more cylinders with inter-connected piston assemblies.

CYLINDER, TELESCOPING: A cylinder with nested multiple tubular rod segments which provide a long working stroke in a short retracted envelope.

DUROMETER HARDNESS: A comparative indication of elastomer hardness determined by a durometer.

ELBOW: A connector that makes an angle between mating lines. The angle is always 90 degrees unless another angle is specified.

FILTER: A device whose primary function is the retention by a porous media of insoluble contaminants from a fluid.

FILTER ELEMENT: The porous device which performs the actual process of filtration.

FILTER MEDIA, DEPTH: Porous materials which primarily retain contaminant within a tortuous path.

FILTER MEDIA, SURFACE: Porous materials which primarily retain contaminants on the influent face.

FITTING, COMPRESSION: A fitting which seals and grips by manual adjustable deformation.

FITTING, FLANGE: A fitting which utilizes a radially extending collar for sealing and connection.

FITTING, FLARED: A fitting which seals and grips by a preformed flare at the end of the tube.

FITTING, FLARELESS: A fitting which seals and grips by means other than a flare.

FITTING, REUSABLE HOSE: A hose fitting that can be removed from a hose and reused.

FLASH POINT: The temperature to which a liquid must be heated under specified conditions of the test method to give off sufficient vapor to form a mixture with air that can be ignited momentarily by a specified flame.

FLOW, LAMINAR: A flow situation in which fluid moves in aprallel lamina or layers.

FLOW, TURBULENT: A flow situation in which the fluid particles move in a random manner.

FLOW RATE: The volume, mass, or weight of a fluid passing through any conductor per unit of time.

FLOWMETER: A device which indicates either flow rate, total flow, or a combination of both.

FLUID: A liquid or a gas.

FLUID, FIRE RESISTANT: A fluid difficult to ignite which shows little tendency to propagate flame.

FLUID POWER: Energy transmitted and controlled through use of a pressurized fluid.

FLUID POWER SYSTEM: A system that transmits and controls power through use of a pressurized fluid within an enclosed circuit.

FLUIDICS: Engineering science pertaining to the use of fluid dynamic phenomina to sense, control, process information, and/or actuate.

GAGE: An instrument or device for measuring, indicating, or comparing a physical characteristic.

GLAND: The cavity of a stuffing box.

HEAD: The cylinder end closure which covers the differential area between the bore area and the piston rod area.

HEAD: The height of a column or body of fluid above a given point expressed in linear units. Head is often used to indicate gage pressure. Pressure is equal to the height times the density of the fluid.

HEAD, STATIC: The height of a column or body of fluid above a given point.

HEAD, VELOCITY: The equivalent head through which the liquid would have to fall to attain a given velocity. Mathematically is is equal to the square of the velocity (in feet) divided by 64.4 feet per second squared. $h=v2/2g$.

HEAT EXCHANGER: A device which transfers heat through a conducting wall from one fluid to another.

HYDRAULICS: Engineering science pertaining to liquid pressure and flow.

HYDRODYNAMICS: Engineering science pertaining to the energy of liquid flow and pressure.

HYDROKINETICS: Engineering science pertaining to the energy of liquids in motion.

HYDROPNEUMATICS: Pertaining to the combination of hydraulic and pneumatic fluid power.

HYDROSTATICS: Engineering science pertaining to the energy of liquids at rest.

INHIBITOR: Any substance which slows or prevents such chemical reactions as corrosion or oxidation.

INTENSIFIER: A device which converts low pressure fluid power into higher pressure fluid power.

JOINT, SWIVEL: A joint which permits variable operational positioning of lines.

LUBRICATOR: A device which adds controlled or metered amounts of lubricant into a fluid power system.

MANIFOLD: A conductor which provices multiple connection ports.

MANOMETER: A differential pressure gage in which pressure is indicated by the height of a liquid column of known density. Pressure is equal to the difference in vertical height between two connected columns multiplied by the density of the manometer liquid. Some forms of manometers are "U" tube, inclined tube, well, and bell types.

MICRON: A millionth of a meter or about 0.00004 inch.

MOTOR: A device which converts fluid power into mechanical force and motion. It usually provides rotary mechanical motion.

MOTOR, FIXED DISPLACEMENT: A motor in which the displacement per unit of output motion cannot be varied.

MOTOR, ROTARY, LIMITED: A rotary motor having limited motion.

MOTOR, VARIABLE DISPLACEMENT: A motor in which the displacement per unit of output motion can be varied.

MUFFLER: A device for reducing gas flow noise. Noise is decreased by back pressury control of gas expansion.

NEUTRALIZATION NUMBER: A measure of the total acidity or bascity of an oil; this includes organic or inorganic acids or bases or a combination thereof (ASTM Designation D974-58T).

NEWT: The standard unit of kinematic viscosity in the English system. It is expressed in square inches per second.

NIPPLE: A short length of pipe or tube.

PACKING: A sealing device consisting of bulk deformable material or one or more mating deformable elements, reshaped by manually adjustable compression to obtain and maintain effectiveness. It usually uses axial compression to obtain radial sealing.

PASCAL'S LAW: A pressure applied to a confined fluid at rest is transmitted with equal intensity throughout the fluid.

PASSAGE: A machined or cored fluid-conducting path which lies within or passes through a component.

PIPE: A line whose outside diameter is standardized for threading. Pipe is available in Standard, Extra Strong,

Double Extra Strong or Schedule wall thicknesses.

PNEUMATICS: Engineering science pertaining to gaseous pressure and flow.

POISE: The standard unit of absolute viscosity in the c.g.s. (centimeter-gram-second) system. It is the ratio of the shearing stress to the shear rate of a fluid and is expressed in dyne seconds per square centimeter; 1 centipoise equals .01 poise.

PORT: An internal or external terminus of a passage in a component.

POUR POINT: The lowest temperature at which a fluid will flow under specified conditions (ASTM Designation D97-57).

POWER UNIT: A combination of pump, pump drive, reservoir, controls and conditioning components which may be required for its application.

PRESSURE: Force per unit area, usually expressed in pounds per square inch.

PRESSURE, ABSOLUTE: The sum of atmospheric and gage pressures.

PRESSURE, ATMOSPHERIC: Pressure exerted by the atmosphere at any specific location. (Sea level pressure is approximately 14.7 pounds per square inch absolute.)

PRESSURE, BACK: The pressure encountered on the return side of a system.

PRESSURE, CRACKING: The pressure at which a pressure operated valve begins to pass fluid.

PRESSURE, DIFFERENTIAL: The difference in pressure between any. two points of a system or a component.

PRESSURE, GAGE: Pressure differential above or below atmospheric pressure.

PRESSURE, MAXIMUM INLET: The maximum rated gage pressure applied to the inlet.

PRESSURE, OPERATING: The pressure at which a system is operated.

PRESSURE, OVERRIDE: The difference between the cracking pressure of a valve and the pressure reached when the valve is passing full flow.

PRESSURE, PEAK: The maximum pressure encountered in the operation of a component.

PRESSURE, PROOF: The non-destructive test pressure in excess of the maximum rated operating pressure.

PRESSURE, RATED: The qualified operating pressure which is recommended for a component or a system by the manufacturer.

PRESSURE, SHOCK: The pressure existing in a wave moving at supersonic velocity.

PRESSURE, SYSTEM: The pressure which overcomes the total resistances in a system. It includes all losses as well as useful work.

PRESSURE-SWITCH: An electric switch operated by fluid pressure.

PRESSURE, WORKING: The pressure which overcomes the resistance of the working device.

PUMP: A device which converts mechanical force and motion into hydraulic fluid power.

PUMP, FIXED DISPLACEMENT: A pump in which the displacement per cycle cannot be varied.

PUMP, VARIABLE DISPLACEMENT: A pump in which the displacement per cycle can be varied.

REDUCER: A connector having a smaller line size at one end than the other.

RESERVOIR: A container for storage of liquid in a fluid power system.

RESTRICTOR: A device which reduces the cross-sectional flow area.

RESTRICTOR, CHOKE: A restrictor, the length of which is relatively large with respect to its cross-sectional area.

RESTRICTOR, ORIFICE: A restrictor, the length of which is relatively small with respect to its cross-sectional area. The orifice may be fixed or variable. Variable types are non-compensated, pressure compensated, or pressure and temperature compensated.

REYN: The standard unit of absolute viscosity in the English system. It is expressed in pound-seconds per square inch.

SEAL, LIP: A sealing device which has a flexible sealing projection.

SEALING DEVICE: A device which prevents or controls the escape of a fluid or entry of a foreign material.

SILENCER: A device for reducing gas flow noise. Noise is decreased by tuned resonant control of gas expansion.

SPECIFIC GRAVITY (LIQUID): The ratio of the weight of a given volume of liquid to the weight of an equal volume of water.

STOKE: The standard unit of kinematic viscosity in the c.g.s. (centimeter-gram-second) system. It is expressed in square centimeters per second; 1 centistoke equals .01 stoke.

STRAINER: A coarse filter.

SURGE: A momentary rise of pressure in a circuit.

TEE: A connector with three ports, a pair on one axis with one side outlet at right angles to this axis.

TIE ROD: An axial external cylinder element which traverses the length of the cylinder. It is pre-stressed at assembly to hold the ends of the cylinder against the tubing. Tie rod extensions can be a mounting device.

TRUNNION: A mounting device consisting of a pair of opposite projecting cylindrical pivots. The cylindrical pivot pins are at right angle or normal to the piston rod centerline to permit the cylinder to swing in a plane.

TUBE: A line whose size is its outside diameter. Tube is available in varied wall thicknesses.

UNION: A connector which permits lines to be joined or separated without requiring the lines to be rotated.

VALVE: A device which controls fluid flow direction, pressure, or flow rate.

VALVE, DIRECTIONAL CONTROL: A valve whose primary function is to direct or prevent flow through selected passages.

VALVE, DIRECTIONAL CONTROL, CHECK: A directional control valve which permits flow of fluid in only one direction.

VALVE, DIRECTIONAL CONTROL, FOUR WAY: A directional control valve whose primary function is to alternately pressurize and exhaust two working ports.

VALVE, DIRECTIONAL CONTROL, SERVO: A directional control valve which modulates flow or pressure as a function of its input signal.

VALVE, DIRECTIONAL CONTROL, STRAIGHTWAY: A two port directional control valve which modulates flow or pressure as a function of its input signal.

VALVE, DIRECTIONAL CONTROL, THREE WAY: A directional control valve whose primary function is to alternately pressurize and exhaust a working port.

VALVE, FLOW CONTROL: A valve whose primary function is to control flow rate.

VALVE, FLOW CONTROL, DECELERATION: A flow control valve which gradually reduces flow rate to provide deceleration.

VALVE, FLOW CONTROL, PRESSURE COMPENSATED: A flow control valve which controls the rate of flow independent of system pressure.

VALVE, FLOW CONTROL, PRESSURE-TEMPERATURE COMPENSATED: A pressure compensated flow control valve which controls the rate of flow independent of fluid temperature.

VALVE, FLOW DIVIDING: A valve which divides the flow from a single source into two or more branches.

VALVE, FLOW DIVIDING, PRESSURE COMPENSATED: A flow dividing valve which divides the flow at constant ratio regardless of the difference in the resistances of the branches.

VALVE, FOUR POSITION: A directional control valve having four positions to give four selections of flow.

VALVE, PREFILL: A valve which permits full flow from a tank to a "working" cylinder during the advance portion of a cycle, permits the operating pressure to be applied to the cylinder during the working portion of the cycle, and permits free flow from the cylinder to the tank during the return portion of the cycle.

VALVE, PRESSURE CONTROL, COUNTERBALANCE: A pressure control valve which maintains back pressure to prevent a load from falling.

VALVE, PRESSURE CONTROL, DE-COMPRESSION: A pressure control valve that controls the rate at which the contained energy of the compressed fluid is released.

VALVE, PRESSURE CONTROL, PRESSURE REDUCING: A pressure control valve whose primary function is to limit outlet pressure.

VALVE, PRESSURE CONTROL, RELIEF: A pressure control valve whose primary function is to limit system pressure.

VALVE, SEQUENCE: A valve whose primary function is to direct flow in a pre-determined sequence.

VALVE, SHUTTLE: A connective valve which selects one of two or more circuits because flow or pressure changes between the circuits.

VALVE, SHUTOFF: A valve which operates fully open or fully closed.

VALVE, THREE POSITION: A directional control valve having three positions to give three selections of flow.

VALVE, TWO POSITION: A directional control valve having two positions to give two selections of flow conditions.

VALVE FLOW CONDITION, CLOSED: All ports are closed.

VALVE FLOW CONDITION, FLOAT: Working ports are connected to exhaust or reservoir.

VALVE FLOW CONDITION, HOLD: Working ports are blocked to hold a powered device in a fixed position.

VALVE FLOW CONDITION, OPEN: All ports are open.

VALVE FLOW CONDITION, TANDEM: Working ports are blocked and supply is connected to the reservoir port.

VALVE MOUNTING, BASE: The valve is mounted to a plate which has top and side ports.

VALVE MOUNTING, LINE: The valve is mounted directly to system lines.

VALVE MOUNTING, MANIFOLD: The valve is mounted to a plate which provides multiple connection ports for two or more valves.

VALVE MOUNTING, SUB-PLATE: The valve is mounted to a plate which provides straight-through top and bottom ports.

VALVE POSITION, DETENT: A predetermined position maintained by a holding device acting on the flow-directing elements of a directional control valve.

VALVE POSITION, NORMAL: The valve position when signal or actuating force is not being applied.

VISCOSITY: A measure of the internal friction or the resistance of a fluid to flow.

VISCOSITY, ABSOLUTE: The ratio of the shearing stress to the shear rate of a fluid. It is usually expressed in centipoise.

VISCOSITY, KINEMATIC: The absolute viscosity divided by the density of the fluid. It is usually expressed in centistokes.

VISCOSITY, SUS: Saybolt Universal Seconds (SUS), which is the time in seconds for 60 milliliters of oil to flow through a standard orifice at a given temperature (ASTM Designation D88-56).

VISCOSITY INDEX: A measure of the viscosity-temperature characteristics of a fluid as referred to that of two arbitrary reference fluids (ASTM Designation D567-53).

WYE (Y): A connector with three ports, a pair on one axis with one side outlet at any angle other than right angles to this axis. The side outlet angle is usually 45°, unless another angle is specified.

Index